Nanostructured Materials and Nanotechnology IV

Nanostructured Materials and Nanotechnology IV

A Collection of Papers Presented at the 34th International Conference on Advanced Ceramics and Composites January 24–29, 2010 Daytona Beach, Florida

Edited by
Sanjay Mathur
Suprakas Sinha Ray

Volume Editors
Sanjay Mathur
Tatsuki Ohji

The American Ceramic Society

A John Wiley & Sons, Inc., Publication

Published by John Wiley & Sons, Inc., Hoboken, New Jersey.
Published simultaneously in Canada.

For general information on our other products and services or for technical support, please contact our Customer Care Department within the United States at (800) 762-2974, outside the United States at (317) 572-3993 or fax (317) 572-4002.

Wiley also publishes its books in a variety of electronic formats. Some content that appears in print may not be available in electronic format. For information about Wiley products, visit our web site at www.wiley.com.

Library of Congress Cataloging-in-Publication Data is available.

ISBN 978-0-470-59472-8

Printed in the United States of America.

10 9 8 7 6 5 4 3 2 1

Contents

Preface

The 4th International Symposium on Nanostructured Materials and Nanotechnology was held during the 34th International Conference and Exposition on Advanced Ceramics and Composites, in Daytona Beach, Florida during January 24–29, 2010. This symposium provided, for the fourth consecutive year, an international forum for scientists, engineers, and technologists to discuss new developments in the field of nanotechnology. This year's symposium had a special focus on the large-scale production and potential of nanomaterials in energy applications. The symposium covered a broad perspective including synthesis, processing, modeling and structure-property correlations in Nanomaterials and nanocomposites. More than 90 contributions (invited talks, oral presentations, and posters) were presented by participants from universities, research institutions, and industry, which offered interdisciplinary discussions indicating strong scientific and technological interest in the field of nanostructured systems. The geographical spread of the symposium was impressive with participants coming from 16 nations.

This issue contains 17 peer-reviewed (invited and contributed) papers covering various aspects and the latest developments related to processing, modeling and manufacturing technologies of nanoscaled materials including inorganic-organic nanocomposites, nanowire-based sensors, new generation photovoltaic cells, self-assembly of nanostructures, functional nanostructures for cell tracking and heterostructures. Each manuscript was peer-reviewed using The American Ceramic Society review process.

The editors wish to extend their gratitude and appreciation to all the authors for their cooperation and contributions, to all the participants and session chairs for their time and efforts, and to all the reviewers for their valuable comments and suggestions. Financial support from the Engineering Ceramic Division of The American Ceramic Society is gratefully acknowledged. The invaluable assistance of the ACerS's staff of the meetings and publication departments, instrumental in the success of the symposium, is gratefully acknowledged,

We believe that this issue will serve as a useful reference for the researchers and technologists interested in science and technology of nanostructured materials and devices.

SANJAY MATHUR
University of Cologne
Cologne, Germany

SUPRAKAS SINHA RAY
National Centre for Nano Structured Materials
CSIR, Pretoria, South Africa

Introduction

This CESP issue represents papers that were submitted and approved for the proceedings of the 34th International Conference on Advanced Ceramics and Composites (ICACC), held January 24–29, 2010 in Daytona Beach, Florida. ICACC is the most prominent international meeting in the area of advanced structural, functional, and nanoscopic ceramics, composites, and other emerging ceramic materials and technologies. This prestigious conference has been organized by The American Ceramic Society's (ACerS) Engineering Ceramics Division (ECD) since 1977.

The conference was organized into the following symposia and focused sessions:

Symposium 1	Mechanical Behavior and Performance of Ceramics and Composites
Symposium 2	Advanced Ceramic Coatings for Structural, Environmental, and Functional Applications
Symposium 3	7th International Symposium on Solid Oxide Fuel Cells (SOFC): Materials, Science, and Technology
Symposium 4	Armor Ceramics
Symposium 5	Next Generation Bioceramics
Symposium 6	International Symposium on Ceramics for Electric Energy Generation, Storage, and Distribution
Symposium 7	4th International Symposium on Nanostructured Materials and Nanocomposites: Development and Applications
Symposium 8	4th International Symposium on Advanced Processing and Manufacturing Technologies (APMT) for Structural and Multifunctional Materials and Systems
Symposium 9	Porous Ceramics: Novel Developments and Applications
Symposium 10	Thermal Management Materials and Technologies
Symposium 11	Advanced Sensor Technology, Developments and Applications

Focused Session 1 Geopolymers and other Inorganic Polymers
Focused Session 2 Global Mineral Resources for Strategic and Emerging
                   Technologies
Focused Session 3 Computational Design, Modeling, Simulation and
                   Characterization of Ceramics and Composites
Focused Session 4 Nanolaminated Ternary Carbides and Nitrides (MAX Phases)

The conference proceedings are published into 9 issues of the 2010 Ceramic Engineering and Science Proceedings (CESP); Volume 31, Issues 2–10, 2010 as outlined below:

- Mechanical Properties and Performance of Engineering Ceramics and Composites V, CESP Volume 31, Issue 2 (includes papers from Symposium 1)
- Advanced Ceramic Coatings and Interfaces V, Volume 31, Issue 3 (includes papers from Symposium 2)
- Advances in Solid Oxide Fuel Cells VI, CESP Volume 31, Issue 4 (includes papers from Symposium 3)
- Advances in Ceramic Armor VI, CESP Volume 31, Issue 5 (includes papers from Symposium 4)
- Advances in Bioceramics and Porous Ceramics III, CESP Volume 31, Issue 6 (includes papers from Symposia 5 and 9)
- Nanostructured Materials and Nanotechnology IV, CESP Volume 31, Issue 7 (includes papers from Symposium 7)
- Advanced Processing and Manufacturing Technologies for Structural and Multifunctional Materials IV, CESP Volume 31, Issue 8 (includes papers from Symposium 8)
- Advanced Materials for Sustainable Developments, CESP Volume 31, Issue 9 (includes papers from Symposia 6, 10, and 11)
- Strategic Materials and Computational Design, CESP Volume 31, Issue 10 (includes papers from Focused Sessions 1, 3 and 4)

The organization of the Daytona Beach meeting and the publication of these proceedings were possible thanks to the professional staff of ACerS and the tireless dedication of many ECD members. We would especially like to express our sincere thanks to the symposia organizers, session chairs, presenters and conference attendees, for their efforts and enthusiastic participation in the vibrant and cutting-edge conference.

ACerS and the ECD invite you to attend the 35th International Conference on Advanced Ceramics and Composites (http://www.ceramics.org/icacc-11) January 23–28, 2011 in Daytona Beach, Florida.

Sanjay Mathur and Tatsuki Ohji, Volume Editors
July 2010

CORE - SHELL NANOSTRUCTURES: SCALABLE, ONE-STEP AEROSOL SYNTHESIS AND IN-SITU SiO₂ COATING AND FUNCTIONALIZATION OF TIO₂ AND Fe₂O₃ NANOPARTICLES

J.T.N. Knijnenburg, A. Teleki, B. Buesser, S.E. Pratsinis
Particle Technology Laboratory, Institute of Process Engineering
Department of Mechanical and Process Engineering
Swiss Federal Institute of Technology, ETH Zurich
Sonneggstrasse 3, CH-8092 Zurich, Switzerland

ABSTRACT

Scalable flame aerosol synthesis of surface-functionalized and coated nanoparticles in one-step is presented. These composite materials provide the functionality of nanoparticles and even "cure" any deleterious effects that might have. So they can be incorporated readily in a liquid or polymer matrix. Magnetic maghemite core nanoparticles are made by a spray flame and coated in-situ by judiciously positioning (as guided by computational fluid dynamics simulations) a hollow ring that delivers in swirling mode precursor vapor for the SiO_2 shell through 1 to 16 jets. This results in hermetically-coated superparamagnetic particles (15-20 nm) with a 1-3 nm thin silica film as determined by microscopy and thermal conductivity measurements. Furthermore, the extent of surface functionalization by organic coatings is evaluated by infrared spectroscopy and dynamic light scattering of particle suspensions in organic solvents.

INTRODUCTION

The need for coating or surface modification of nanostructured particles comes from the fact that typically the surface properties of as-prepared particles are different than the desired ones for the final applications. In such systems, the core particle possesses functionality while the shell enhances or facilitates the interaction with the host liquid or solid matrix. A classic example of functional core particles is the coating of TiO_2 pigment particles with SiO_2. Light scattering by the TiO_2 core gives the desired white color whereas the SiO_2 shell allows incorporating the particles in paints or polymers by blocking the reactivity of TiO_2 which would otherwise degrade the surrounding material.[1] In other applications (e.g. V_2O_5-TiO_2 catalysts) the core provides the support (TiO_2) of the catalytically active V_2O_5 for deNOx selective catalytic reduction[2] or for synthesis of phthalic anhydride.[3]

Also magnetic nanoparticles can bind to drugs and proteins that can be guided in the human body using external magnetic fields.[4] Although metal oxides have a lower magnetization than metallic nanoparticles, the latter are highly reactive and toxic and therefore less suitable for biomedical applications.[5] Iron oxide nanoparticles have been widely investigated for in vivo applications (e.g. magnetic resonance imaging (MRI) contrast enhancement and drug delivery).[4,5] Such magnetic iron oxide nanoparticles are frequently coated with silica to improve their functionality and biocompatibility.[5] The silica coating makes the nanoparticles stable in aqueous conditions and limits magnetically induced self-agglomeration of magnetic cores. Furthermore, the silanol groups on the silica surface are able to react further with alcohols and silanes to make stable nonaqueous suspensions and can be further modified by covalent bonding of specific ligands.[4] Moreover, silica-coated or -embedded maghemite (γ-Fe_2O_3) nanoparticles exhibit improved thermal stability. Pure or uncoated γ-Fe_2O_3 is thermally unstable and is transformed to hematite (α-Fe_2O_3), the most stable form at high temperatures.[6] Silica inhibits such transformations of γ-Fe_2O_3 in O_2 or in air.[7,8] Various studies have been done on the incorporation of Fe_xO_y nanoparticles in polymers to obtain superparamagnetic, transparent nanocomposites.[9,10] Coating maghemite with silica facilitates homogeneous distribution of nanoparticles in a polymer matrix.[11] Typically such coatings are made by a sol-gel process[7,12] involving several steps, as Fe_xO_y core formation and its SiO_2 coating are two separate unit operations.

1

Nanocomposites combine the advantages of the inorganic filler material (e.g. thermal stability, rigidity) and the organic host polymer (e.g. flexibility, processability), having enhanced electrical, optical or mechanical properties over the individual components.[13,14] Surface functionalization of hydrophilic nanoparticles increases their hydrophobicity and can also include reactive groups (e.g. vinyl groups) that can cause further crosslinking between particles and host polymer.[14] For example, silica dispersibility into an acrylic-based polyurethane is improved by surface functionalization of hydrophilic silica with long chain coupling agents like octyltriethoxysilane (OTES) while the non-functionalized silica showed inhomogeneous dispersion that reduces strength and transparency of the nanocomposite.[15]

Wet phase functionalization of flame-made particles has been studied in some detail. For example, flame-made radiopaque[16] and amorphous Ta_2O_5/SiO_2 nanoparticles were functionalized[17] and subsequently incorporated into a polyacrylate matrix.[16,18] Functionalization of these powders facilitated dispersion into the polymer matrix.[18]

Particle suspensions are generally sterically stabilized with an organic adlayer that acts as a steric barrier counteracting the attractive forces causing agglomeration.[19] Long shelf-life suspensions have direct applications as electronic ink displays, where the image quality is affected by the uniformity and stability of particle dispersions inside an organic medium.[20,21] Dilute suspensions are used for biomedical applications,[4,5] while the more concentrated ones (30-40 vol%) find their application in further processing into organic films on devices (i.e. sensors or solar cells) by screen-printing or tape casting.[22,23]

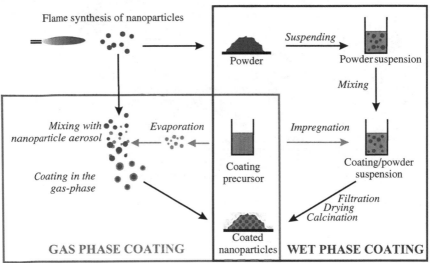

Figure 1: Comparison of conventional wet-phase and in-situ flame coating methods. In-situ flame coating allows preparation in one step, whereas the conventional route is a multi-step process.

Typically[3] surface functionalization processes have multiple steps (ex-situ synthesis) where the core material is formed first, followed by a second operation where coating is done (Figure 1: wet phase coating).[24] In-situ processes are available as well in the wet phase, where the coating is done at the same time as core particle is formed.[15,24,25] There is, however, an increasing interest in rapid gas phase processes. Aerosol routes are attractive as they offer fewer process steps than conventional wet chemistry, easier particle collection from gaseous rather than liquid streams and no liquid by-products that require costly cleaning.[26] In addition, aerosol-made particles and films have unique morphology and high purity (e.g. optical fibers) and sometimes even metastable phases[27] (e.g. low temperature $BaCO_3$ for NOx storage-reduction or ε-WO_3 for ultraselective sensing of acetone, a tracer of diabetes[28]) leading to synthesis of mixed oxides, metal salts and pure metals as well as layered particles and solid or highly porous films with unique functionality resulting in novel catalysts, micropatterned sensors, phosphors, battery, solar and fuel cell electrodes, dental prosthetics, nutritional supplements[27] and even durable sorbents for CO_2 sequestration.[29]

Many metal oxides (e.g. TiO_2, SiO_2) are made today on a large scale in flame aerosol reactors. However for the coating of these products often costly and extensive wet-phase processes are required. In fact, the cost of coating pigmentary TiO_2 particles is comparable to their coating and functionalization cost.[30] As a result, there is a need for continuous gas phase coating or functionalization processes and in particular to simplify and improve such processes by combining both particle synthesis and coating in a one-step gas phase process (Figure 1: gas phase coating). Coherent and homogeneous coating on all particles is essential for optimal performance. A "coherent coating" in this context means that the core particle and the coating material stick (cohere) together. A "homogeneous coating" signifies that the coating has the same thickness around the core particle, i.e. the silica is distributed evenly over the core surface without any "patches". Furthermore, optimization of coating thickness is desired, as silica decreases the UV absorbance of TiO_2 (e.g. in sunscreens) and increases production costs. Scalable production of pure[31] and mixed oxides[32] up to 1 kg/h can be achieved using flame spray aerosol processes even in academic laboratories. At industrial facilities this is in the order of 25 tons/h.

IN-SITU COATING OF FLAME-MADE TITANIA

An alternative process for the scalable manufacture of silica-coated nanoparticles in one step is flame spray pyrolysis (FSP).[33-35] Various studies have been performed on coating of freshly-made TiO_2 with silica. Teleki et al.[34] coated rutile TiO_2 particles by injection of hexamethyldisiloxane (HMDSO), a Si precursor, downstream of the TiO_2 formation zone in an enclosed FSP reactor. Figure 2a shows the experimental setup, where the FSP reactor is enclosed by a 5-30 cm long quartz glass tube (4.5 cm ID).

There Al-doped (4 wt% Al_2O_3) rutile TiO_2 particles were produced by FSP of a 1 M solution of aluminum sec-butoxide and titanium-tetra-isopropoxide in xylene. The precursor solution was injected through the inner capillary with a flow rate of 5 mL/min, dispersed with 5 L/min O_2 (pressure drop 1.5 bar) and sheathed by another 40 L/min O_2. The solution spray was ignited by a ring shaped methane/oxygen (1.5/3.2 L/min) premixed flame.[33,36] At the top of the lower glass tube, a stainless steel metal torus pipe (0.38 cm ID) ring (4.5 cm ID) with 1, 2, 4, 8 or 16 radial outlets (0.06 cm each in diameter) was placed. The outlets are directed 10° away from the ring radius and pointed 20° downstream[37] to enable judicious swirl mixing of the coating precursor vapor with the titania-containing aerosol stream. Above the ring, another 30 cm long quartz glass tube (4.5 cm ID) was placed. Through the torus ring outlets, a gas flow of 0.8 L/min N_2 carrying HMDSO vapor (the Si coating precursor) was injected with an additional 15 L/min N_2. The product particles are collected on a glass fiber filter using a vacuum pump.

The effect of burner-ring-distance (BRD) on the coating quality was determined experimentally.[34] Injection of HMDSO vapor at low BRD (where high temperature and TiO_2 formation still takes place) leads to fast oxidation of HMDSO, resulting in separate SiO_2 and TiO_2 particles or domains. In contrast, all particles were coated homogeneously with 2-3 nm thick SiO_2 films and no separate domains or particles were obtained when HMDSO was injected at BRD ≥ 20 cm. The coating quality was verified by photooxidation of isopropyl alcohol (IPA).[38] Titania is photocatalytically active converting IPA to acetone. So the released acetone concentration by photooxidation of IPA slurries containing SiO₂-coated TiO₂ made at various BRD was determined by gas conductivity measurements. This confirmed the formation of hermetic SiO₂ coatings onto the TiO₂ particles.[35]

The influence of silica content on the coating morphology was studied at a BRD of 20 cm.[34] For 5 wt% SiO₂ no coating film was visible by TEM due to the low theoretical coating thickness (< 1 nm for a 40 nm particle). The existence of such SiO₂ shells was confirmed[34] as it had reduced the photooxidation of IPA to acetone to 50%. Increasing to 10 or 20 wt% SiO₂ yielded homogeneous hermetic coatings of 2-4 nm thick in agreement with the expected theoretical coating thickness. For these concentrations, no separate silica particles were observed by TEM while the quality of the coating was confirmed by photocatalytic evaluation of such particle slurries containing IPA.

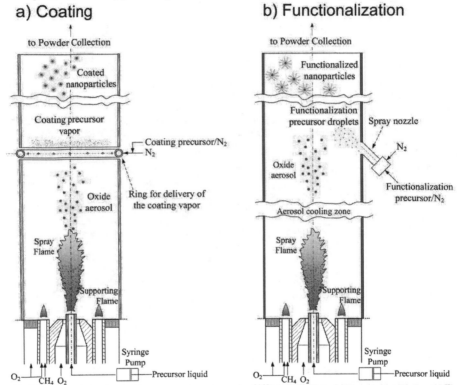

Figure 2: Experimental setup for (a) in-situ SiO₂ coating[34,39] and (b) in-situ surface functionalization[40] of flame-made TiO₂ or Fe₂O₃ particles.

The mixing of aerosol with vapor during TiO₂ coating by SiO₂ was studied also by varying N₂ flow rate and the number of coating jets. Such coating was optimized by systematic design using computational fluid dynamics (CFD) where the mixing of the incoming HDMSO-containing N₂ flow with the product TiO_2 aerosol was visualized. The volume fraction of the carrier N_2 in the stream is an indication for the mixing efficiency of HMDSO as N_2 is the main component of the coating precursor stream. Figure 3 shows the N_2 volume fraction distribution along the tube axis above the ring for various N_2 volume flow rates. The flows at the jet outlets are red as they contain 100 vol% N_2. At a low overall jet flow rate of 5 L/min (Figure 3a) the N_2 flow remains largely near the tube wall while N_2-poor regions remain along the centerline of the tube. Increasing the flow rate to 10 L/min (Figure 3b) improves homogeneity along and across the tube, however still rather limited mixing of the HMDSO-laden vapor stream with the TiO_2 particle-containing aerosol takes place. For high flow rates (15 L/min), some N_2 lean regions (blue) are still present above the torus ring along the reactor walls by the high jet exit velocity,[33] but there is already improved mixing at 3 cm above the injection point, shown by the rather homogeneous concentration of 15-30 vol% N_2 (green). With increasing height above the torus ring, however, the flow becomes more homogeneous throughout the tube,[33] in stark contrast to the lower jet flow rates (Figure 3a,b).

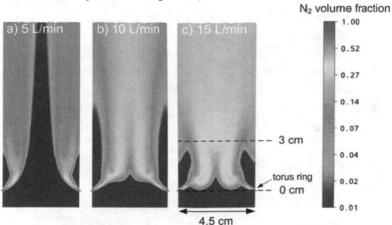

Figure 3: Contours of the N_2 volume fraction as visualized by CFD simulations issuing from 16 jets of a torus ring[34] with a volume flow of 5 (a), 10 (b) and 15 L/min N_2 (c, adopted from Teleki et al.[33]). The logarithmic color scale ranges from < 0.01 (blue) to 1 (red) N_2.

The homogeneity of the SiO_2 coating on the rutile TiO_2 nanoparticles was verified experimentally at different HMDSO-laden N_2 injection flow rates: Figure 4 displays TEM images of rutile TiO_2 nanoparticles coated with 20 wt% SiO_2 at various N_2 flow rates. At 5 (Figure 4a) and 10 (Figure 4b) L/min, the coating is rather inhomogeneous and large domains of separate amorphous SiO_2 particles are clearly visible. For 20 (Figure 4c) and 30 (Figure 4d) L/min N_2, all titania particles appear to be coated without any separate SiO_2 particles. This is attributed to the improved mixing and the decreased silica precursor concentration which forms smaller silica particles that can deposit and sinter faster on the titania rather than collide with other SiO_2 coating particles and form larger ones. As a result, smooth coating films are formed.

Teleki et al.[33] have visualized by CFD the effect of the number of jets issuing from the torus pipe. This plays a large role in the homogeneity of the flow. For increasing number of jet outlets, the N_2 concentration is more homogeneous and the mixing quality increases. For only 1 jet, the N_2 concentration was enriched at the walls opposite of the coating precursor inlet. With 2 or 4 inlets still poorly mixed regions exist, while with 16 jets the concentration profile is uniform over the tube radius. These results were verified also experimentally: Separate silica domains as well as uncoated and homogeneously SiO_2-coated TiO_2 particles were observed with 1 and 4 inlets with low N_2 flows; for 8 inlets, at low injection velocity still separate silica is observed, however increasing the flow rate has shown to improve the coating quality and efficiency.

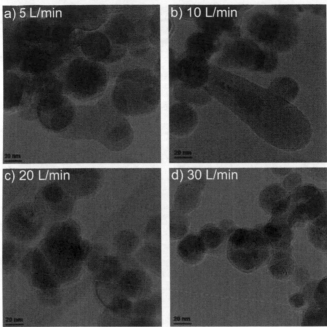

Figure 4: TEM images of 20 wt% SiO_2-coated rutile TiO_2 particles produced with 5 (a), 10 (b), 20 (c) and 30 (d) L/min injection flow rate of coating precursor HMDSO - N_2 stream.

SILICA COATING OF SUPERPARAMAGNETIC IRON OXIDE NANOPARTICLES

The potential of the above process involving sequential core oxide particle formation followed by coating it with nanothin silica layer is explored in synthesis of iron oxide - silica core shell nanoparticles. So, silica-coated γ-Fe_2O_3 nanoparticles were also made in one step using an enclosed FSP reactor.[39] In this process, the precursor consists of iron(III)acetylacetonate (Fe(acac)$_3$) dissolved in xylene/acetylene (3:1 in volume) to form a 0.34 M solution. The freshly formed Fe_2O_3 nanoparticle aerosol is mixed downstream with HMDSO vapor (Figure 2a).

Figure 5 displays images of pure (a) and SiO_2-coated (b-d) γ-Fe_2O_3 nanoparticles. The uncoated maghemite particles are pure, mostly hexagonally-shaped crystals as typically observed with flame-made iron oxide nanoparticles.[41] At 6.5 wt% SiO_2 coating (Figure 5b) no clear silica layer around the iron oxide could be observed. This concentration would correspond to a theoretical coating thickness of < 1 nm. At 17 (Figure 5c) and 23 wt% SiO_2 (Figure 5d) a homogeneous amorphous film of around 2 nm thick on the γ-Fe_2O_3 nanoparticles is formed. The latter particles are similar to those obtained by sol-gel coating of maghemite with 43 wt % SiO_2.[42] No clear difference in aggregation or agglomeration can be distinguished between the uncoated and coated Fe_2O_3 particles. As confirmed by dynamic light scattering measurements,[39] the SiO_2-coated Fe_2O_3 particles exhibited excellent dispersibility in water compared to that of flame-made co-oxidized SiO_2/Fe_2O_3 and uncoated Fe_2O_3. The presence of the silica coatings decreases the isoelectric point of aqueous suspensions of such nanoparticles from around pH 7 for uncoated Fe_2O_3 to around pH 1.7 for 23 wt% SiO_2-coated Fe_2O_3, thereby inhibiting their magnetically-induced self-agglomeration.[39]

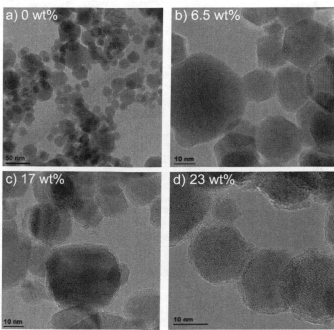

Figure 5: TEM images of flame-made uncoated γ-Fe_2O_3 (a) and γ-Fe_2O_3 nanoparticles coated with 6.5 wt% (b), 17 wt% (c) and 23 wt% (d) SiO_2 by injection of HMDSO vapor.

To compare, mixed SiO_2/Fe_2O_3 particles were also made by FSP of xylene/acetonitrile (75/25 by volume) solutions containing both HMDSO and $Fe(acac)_3$ in the enclosed reactor. In such SiO_2/Fe_2O_3 particles containing 36 (Figure 6a) and 46 wt% SiO_2 (Figure 6b), crystalline iron oxide particles are segregated to the edge of amorphous SiO_2 particles,[42] in contrast to the SiO_2-coated Fe_2O_3 made by introducing the Si-precursor downstream of the Fe_xO_y formation (e.g. Figure 5c,d).

The magnetic properties of the SiO_2-coated γ-Fe_2O_3 nanoparticles were measured by a vibrating sample magnetometer. The powders showed near zero hysteresis and were found to have a slightly lower magnetization than that of pure γ-Fe_2O_3. Their magnetization however was superior to that of co-oxidized SiO_2/Fe_2O_3 and commercial MagSilica. This means that SiO_2-coated γ-Fe_2O_3 nanoparticles retain most of the magnetic properties of Fe_2O_3.

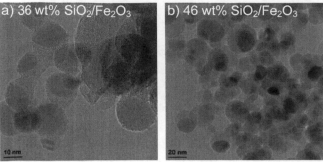

Figure 6: TEM images of co-oxidized 36 (a) and 46 wt% SiO_2/Fe_2O_3 (b) made by flame spray pyrolysis (FSP) of the corresponding precursor solution. These particles contain segregated domains of SiO_2 and maghemite Fe_2O_3 in contrast to coated maghemite with nanothin silica films (Figure 5c,d).

IN-SITU SURFACE MODIFICATION OF NANOPARTICLES

As was described above, recently photocatalytically active[34] TiO_2 and superparamagnetic[39] Fe_2O_3 particles have been prepared and coated in-situ by hermetic, nanothin SiO_2 layers by applying a scalable[31] flame aerosol technology. Here, this process is further developed for continuous surface functionalization of freshly-formed flame-made nanoparticles with organic materials that are typically used to functionalize oxide nanoparticles and facilitate their incorporation in host liquid or solid matrices. The ability of using such liquid coating precursors allows a complete new range of materials to be used. Formation of titania by FSP and sequentially in-situ functionalized with octyl triethoxy silane (OTES)[40], a silane consisting of a hydrophilic head with a hydrophobic octyl tail, is discussed here. The titania is functionalized by the formation of a covalent bond between the silanol and the titania surface, via interaction with OH groups present on the particle surface.[43]

Figure 2b shows the experimental setup for direct and in-situ organic surface functionalization of flame-made particles. Titania (TiO_2) particles are prepared by flame spray pyrolysis[23] of titanium tetraisopropoxide (TTIP) 1M in 50/50 by volume of acetic acid and 2-ethylhexanoic acid. This precursor is fed at 5 ml/min to the burner, dispersed by 5 l/min O_2, and ignited by a premixed flame of 1.5 l/min CH_4 and 3.2 l/min O_2 supplied uniformly through a ring of 6 mm radius around the nozzle. Through the outermost sinter metal plate (inner radius 9 mm) of the FSP burner, 35 l/min O_2 sheath gas is supplied.

Rapid cooling of the aerosol is necessary to allow the coating reaction to take place at the optimal reaction temperature.[40] This is done by injection of high pressure air with flow rate q_1 through a ring with multiple outlets placed 11.5 cm above the burner, similar to the ring for injection of HMDSO (Figure 2a). Further aerosol cooling is provided by a double-walled steel tube through which cooling water is supplied. Higher upstream, a second quench ring is placed through which air is injected with flow rate q_2. Upstream of the cooling zone, a solution of OTES in demineralized water/ethanol is dispersed by N_2 and injected through a nozzle pointing 40° in the flow direction of the system, resulting in coating precursor injection 58 cm above the burner. The OTES concentration was varied in the range of 0.016-0.8M. In comparison, a post-functionalization process was done by first having a titania-producing flame while injecting an OTES-free water/ethanol stream, while in a second step a particle-free solution was sprayed while injecting the OTES-containing coating precursor solution (0.08M).

The precursor solution for surface functionalization is 0.016-0.4 M octyltriethoxysilane (OTES) in distilled water (10 vol%) and ethanol. Pure, hydrophilic anatase TiO_2 particles are prepared by FSP[44] when spraying 2 ml/min of a pure ethanol/water (90/10 by volume) mixture (without OTES) through the functionalization nozzle. Cooling in the reactor is varied at constant OTES concentration (0.08 M) in the sprayed solution. In the following, low cooling rate protocol (Q_L) refers to $q_1 = 0$ l/min and $q_2 = 30$ l/min and high cooling rate protocol (Q_H) refers to $q_1 = 45$ l/min and $q_2 = 30$ l/min with water cooling.[40] Product particles are denoted $yxTiO_2$ where y is L or H for Q_L or Q_H, respectively, and x is the OTES concentration in the precursor solution fed through the functionalization nozzle.

As described by Teleki et al,[40] the steady-state temperature (after 12 minutes of operation) was measured at various positions in the reactor for Q_L and Q_H. The temperature just after the coating precursor injection was around 250 °C for both cooling rates, resulting in a temperature at the functionalization spray between 270 and 370 °C for respectively Q_H and Q_L,[40] which is similar to the temperature used for silylation of silica in aerosol spray[45] or vapor[43] reactors. The residence time, however, in the system applied here is shorter (less than one second).

The specific surface area (SSA) of FSP-made pure and OTES-functionalized TiO_2 was found to increase by injecting air at q_1, since the temperature close to the particle formation zone in the flame is lowered, resulting in smaller particles, due to a reduction in residence time. This was consistent with a smaller crystallite size from XRD at these conditions. Furthermore, all particles were anatase, indicating that the temperature at Q_L was insufficient to form rutile, similar to open[44] FSP synthesis of TiO_2.

The amount of OTES on the particles was verified by TGA-MS. A linear relation was found between the SSA and the TGA mass loss (R = 0.82). For L0.08TiO_2 a lower mass loss was found than for H0.08TiO_2, consistent with the corresponding SSA of the two materials. This can be explained by the fact that the particles produced at Q_H have a higher SSA. Therefore these particles are able to react with a larger amount of OTES.[17]

Figure 7: Images of commercial T805 and in-situ functionalized H0.08TiO_2 suspended in 2-ethylhexanoic acid after one hour (a) and 45 days (b) after ultrasonication.

Figure 7 shows images of suspensions of commercial T805 and H0.08TiO_2 in 2-ethylhexanoic acid (2-EHA). Comparing the samples 1 hour (Figure 7a) and 45 days (Figure 7b) after ultrasonication, part of the T805 particles has settled after 45 days, as a layer of clear solution is visible at the top. After 45 days only a thin clear solution layer was formed on top of the suspension of functionalized FSP-made TiO_2 in 2-EHA here (Figure 7b).

In comparison, Teleki et al.[40] prepared such suspensions using xylene as well. Xylene is more hydrophobic than 2-EHA, and although 2-EHA has a polarizable acid group, it is still hydrophobic and negligibly soluble in water. Suspensions of T805 particles in xylene were found to have settled completely after 5 days, while the suspension of functionalized FSP-made TiO_2 in xylene was still stable after 45 days, demonstrating its improved stability over T805.[40]

For the stabilization of nanoparticles in organic suspensions, the organic tail on the silanol requires to be sufficiently long in order to repulse neighboring nanoparticles.[15] Furthermore the bond formed between OTES and the particle surface should be sufficiently strong, i.e. stable against washing. Therefore both tail length (coating thickness) and layer stability play a role in the sterical stabilization of the suspensions.

CONCLUSIONS

Coating of flame-made nanostructured particles was done in one step by a scalable aerosol process. By injection of HMDSO vapor through a torus ring and mixing with the freshly-made nanoparticle-containing TiO_2 or Fe_2O_3 aerosol, nanothin amorphous silica coatings are formed in-situ on these particles. Early injection of HMDSO vapor caused the formation of segregated SiO_2 domains and particles with Fe_2O_3 or TiO_2 while the coating homogeneity increased with increasing the deposition height above the burner. The coating precursor injection velocity and number of torus ring outlets for vapor injection were optimized by computational fluid dynamic (CFD) simulations. So a higher injection flow rate and an increased number of jets (outlets) resulted in a more homogeneous HMDSO concentration and faster mixing that led to homogeneous coatings. Superparamagnetic iron oxide was hermetically coated with silica, having magnetic properties superior to commercial materials. This was verified by gas conductivity measurements monitoring the desorption of isopropyl alcohol from these particles at elevated temperatures. The process concept was extended also to in-situ surface modification (or functionalization) of titania using liquid phase OTES, which has shown to produce stable particles that were resistant against acetone washing. Particle suspensions of these hydrophobic titania particles produced stable suspensions in hydrophobic media comparable to commercial material. Further development of this process shows a promising outlook and opens a wide new range of coating materials to be used.

ACKNOWLEDGEMENTS
We thank Dr. Frank Krumeich from the EMEZ (Electron Microscopy ETH Zurich) for the TEM. Support by the Swiss National Science Foundation (SNF) grant # 200021-119946/1 is gratefully acknowledged.

REFERENCES
[1]J. H. Braun, Titanium dioxide - A review, *Journal of Coatings Technology,* **69**, 59-72 (1997).
[2]W. J. Stark, K. Wegner, S. E. Pratsinis, and A. Baiker, Flame aerosol synthesis of vanadia-titania nanoparticles: Structural and catalytic properties in the selective catalytic reduction of NO by NH_3, *Journal of Catalysis,* **197**, 182-191 (2001).
[3]B. Schimmoeller, H. Schulz, S. E. Pratsinis, A. Bareiss, A. Reitzmann, and B. Kraushaar-Czarnetzki, Ceramic foams directly-coated with fame-made V_2O_5/TiO_2 for synthesis of phthalic anhydride, *Journal of Catalysis,* **243**, 82-92 (2006).
[4]A. K. Gupta, and M. Gupta, Synthesis and surface engineering of iron oxide nanoparticles for biomedical applications, *Biomaterials,* **26**, 3995-4021 (2005).
[5]A. H. Lu, E. L. Salabas, and F. Schuth, Magnetic nanoparticles: Synthesis, protection, functionalization, and application, *Angewandte Chemie-International Edition,* **46**, 1222-1244 (2007).

[6]R. Zboril, M. Mashlan, and D. Petridis, Iron(III) oxides from thermal processes-synthesis, structural and magnetic properties, Mossbauer spectroscopy characterization, and applications, *Chemistry of Materials,* **14**, 969-982 (2002).

[7]L. Zhang, G. C. Papaefthymiou, and J. Y. Ying, Size quantization and interfacial effects on a novel γ-Fe_2O_3/SiO_2 magnetic nanocomposite via sol-gel matrix-mediated synthesis, *Journal of Applied Physics,* **81**, 6892-6900 (1997).

[8]P. P. C. Sartoratto, K. L. Caiado, R. C. Pedroza, S. W. da Silva, and P. C. Morais, The thermal stability of maghemite-silica nanocomposites: An investigation using X-ray diffraction and Raman spectroscopy, *Journal of Alloys and Compounds,* **434**, 650-654 (2007).

[9]R. F. Ziolo, E. P. Giannelis, B. A. Weinstein, M. P. Ohoro, B. N. Ganguly, V. Mehrotra, M. W. Russell, and D. R. Huffman, Matrix-Mediated Synthesis of Nanocrystalline γ-Fe_2O_3 - a New Optically Transparent Magnetic Material, *Science,* **257**, 219-223 (1992).

[10]B. H. Sohn, and R. E. Cohen, Processible optically transparent block copolymer films containing superparamagnetic iron oxide nanoclusters, *Chemistry of Materials,* **9**, 264-269 (1997).

[11]R. Mohr, K. Kratz, T. Weigel, M. Lucka-Gabor, M. Moneke, and A. Lendlein, Initiation of shape-memory effect by inductive heating of magnetic nanoparticles in thermoplastic polymers, *Proceedings of the National Academy of Sciences of the United States of America,* **103**, 3540-3545 (2006).

[12]Y. Yonemochi, M. Iijima, M. Tsukada, H. Jiang, E. I. Kauppinen, M. Kimata, M. Hasegawa, and H. Kamiya, Microstructure of iron particles reduced from silica-coated hematite in hydrogen, *Advanced Powder Technology,* **16**, 621-637 (2005).

[13]A. C. Balazs, T. Emrick, and T. P. Russell, Nanoparticle polymer composites: Where two small worlds meet, *Science,* **314**, 1107-1110 (2006).

[14]H. Zou, S. S. Wu, and J. Shen, Polymer/silica nanocomposites: Preparation, characterization, properties, and applications, *Chemical Reviews,* **108**, 3893-3957 (2008).

[15]G. D. Chen, S. X. Zhou, G. X. Gu, H. H. Yang, and L. M. Wu, Effects of surface properties of colloidal silica particles on redispersibility and properties of acrylic-based polyurethane/silica composites, *Journal of Colloid and Interface Science,* **281**, 339-350 (2005).

[16]H. Schulz, L. Madler, S. E. Pratsinis, P. Burtscher, and N. Moszner, Transparent nanocomposites of radiopaque, flame-made Ta_2O_5/SiO_2 particles in an acrylic matrix, *Advanced Functional Materials,* **15**, 830-837 (2005).

[17]H. Schulz, S. E. Pratsinis, H. Rugger, J. Zimmermann, S. Klapdohr, and U. Salz, Surface functionalization of radiopaque Ta_2O_5/SiO_2, *Colloids and Surfaces A: Physicochemical and Engineering Aspects,* **315**, 79-88 (2008).

[18]H. Schulz, B. Schimmoeller, S. E. Pratsinis, U. Salz, and T. Bock, Radiopaque dental adhesives: Dispersion of flame-made Ta_2O_5/SiO_2 nanoparticles in methacrylic matrices, *Journal of Dentistry,* **36**, 579-587 (2008).

[19]A. R. Studart, E. Amstad, and L. J. Gauckler, Colloidal stabilization of nanoparticles in concentrated suspensions, *Langmuir,* **23**, 1081-1090 (2007).

[20]B. Comiskey, J. D. Albert, H. Yoshizawa, and J. Jacobson, An electrophoretic ink for all-printed reflective electronic displays, *Nature,* **394**, 253-255 (1998).

[21]P. Murau, and B. Singer, Understanding and Elimination of Some Suspension Instabilities in an Electrophoretic Display, *Journal of Applied Physics,* **49**, 4820-4829 (1978).

[22]K. Hara, M. Kurashige, Y. Dan-oh, C. Kasada, A. Shinpo, S. Suga, K. Sayama, and H. Arakawa, Design of new coumarin dyes having thiophene moieties for highly efficient organic-dye-sensitized solar cells, *New Journal of Chemistry,* **27**, 783-785 (2003).

[23]A. Teleki, N. Bjelobrk, and S. E. Pratsinis, Flame-made Nb- and Cu-doped TiO_2 sensors for CO and ethanol, *Sensors and Actuators B-Chemical,* **130**, 449-457 (2008).

[24]C. L. Lu, and B. Yang, High refractive index organic-inorganic nanocomposites: design, synthesis and application, *Journal of Materials Chemistry*, **19**, 2884-2901 (2009).

[25]V. Khrenov, M. Klapper, M. Koch, and K. Mullen, Surface functionalized ZnO particles designed for the use in transparent nanocomposites, *Macromolecular Chemistry and Physics*, **206**, 95-101 (2005).

[26]S. E. Pratsinis, and S. V. R. Mastrangelo, Material Synthesis in Aerosol Reactors, *Chemical Engineering Progress*, **85**, 62-66 (1989).

[27]R. Strobel, and S. E. Pratsinis, Flame aerosol synthesis of smart nanostructured materials, *Journal of Materials Chemistry*, **17**, 4743-4756 (2007).

[28]L. Wang, A. Teleki, S. E. Pratsinis, and P. I. Gouma, Ferroelectric WO_3 nanoparticles for acetone selective detection, *Chemistry of Materials*, **20**, 4794-4796 (2008).

[29]H. Lu, P. G. Smirniotis, F. O. Ernst, and S. E. Pratsinis, Nanostructured Ca-based sorbents with high $CO2$ uptake efficiency, *Chemical Engineering Science*, **64**, 1936-1943 (2009).

[30]D. M. King, X. H. Liang, B. B. Burton, M. K. Akhtar, and A. W. Weimer, Passivation of pigment-grade TiO_2 particles by nanothick atomic layer deposited SiO_2 films, *Nanotechnology*, **19**, (2008).

[31]R. Mueller, L. Madler, and S. E. Pratsinis, Nanoparticle synthesis at high production rates by flame spray pyrolysis, *Chemical Engineering Science*, **58**, 1969-1976 (2003).

[32]R. Jossen, R. Mueller, S. E. Pratsinis, M. Watson, and M. K. Akhtar, Morphology and composition of spray-flame-made yttria-stabilized zirconia nanoparticles, *Nanotechnology*, **16**, S609-S617 (2005).

[33]A. Teleki, B. Buesser, M. C. Heine, F. Krumeich, M. K. Akhtar, and S. E. Pratsinis, Role of Gas-Aerosol Mixing during in Situ Coating of Flame-Made Titania Particles, *Industrial & Engineering Chemistry Research*, **48**, 85-92 (2009).

[34]A. Teleki, M. C. Heine, F. Krumeich, M. K. Akhtar, and S. E. Pratsinis, In Situ Coating of Flame-Made TiO_2 Particles with Nanothin SiO_2 Films, *Langmuir*, **24**, 12553-12558 (2008).

[35]A. Teleki, M. K. Akhtar, and S. E. Pratsinis, The quality of SiO_2-coatings on flame-made TiO_2-based nanoparticles, *Journal of Materials Chemistry*, **18**, 3547-3555 (2008).

[36]L. Madler, W. J. Stark, and S. E. Pratsinis, Flame-made ceria nanoparticles, *Journal of Materials Research*, **17**, 1356-1362 (2002).

[37]J. P. Hansen, J. R. Jensen, H. Livbjerg, and T. Johannessen, Synthesis of ZnO particles in a quench-cooled flame reactor, *AIChE Journal*, **47**, 2413-2418 (2001).

[38]R. B. Cundall, R. Rudham, and M. S. Salim, Photocatalytic Oxidation of Propan-2-Ol in Liquid-Phase by Rutile, *Journal of the Chemical Society-Faraday Transactions I*, **72**, 1642-1651 (1976).

[39]A. Teleki, M. Suter, P. R. Kidambi, O. Ergeneman, F. Krumeich, B. J. Nelson, and S. E. Pratsinis, Hermetically Coated Superparamagnetic Fe_2O_3 Particles with SiO_2 Nanofilms, *Chemistry of Materials*, **21**, 2094-2100 (2009).

[40]A. Teleki, N. Bjelobrk, and S. E. Pratsinis, Continuous surface functionalization of flame-made TiO_2 nanoparticles, *Langmuir*, in press (2009) DOI: 10.1021/la9037149.

[41]R. Strobel, and S. E. Pratsinis, Direct synthesis of maghemite, magnetite and wustite nanoparticles by flame spray pyrolysis, *Advanced Powder Technology*, **20**, 190-194 (2009).

[42]D. Li, W. Y. Teoh, C. Selomulya, R. C. Woodward, R. Amal, and B. Rosche, Flame-sprayed superparamagnetic bare and silica-coated maghemite nanoparticles: Synthesis, characterization, and protein adsorption-desorption, *Chemistry of Materials*, **18**, 6403-6413 (2006).

[43]R. R. Sever, R. Alcala, J. A. Dumesic, and T. W. Root, Vapor-phase silylation of MCM-41 and Ti-MCM-41, *Microporous and Mesoporous Materials*, **66**, 53-67 (2003).

[44]A. Teleki, S. E. Pratsinis, K. Kalyanasundaram, and P. I. Gouma, Sensing of organic vapors by flame-made TiO_2 nanoparticles, *Sensors and Actuators B-Chemical*, **119**, 683-690 (2006).

[45]Y. S. Chung, S. A. Song, and S. B. Park, Hydrophobic modification of silica nanoparticle by using aerosol spray reactor, *Colloids and Surfaces a-Physicochemical and Engineering Aspects*, **236**, 73-79 (2004).

SELF-ASSEMBLY OF METAL OXIDES - LIQUID PHASE CRYSTAL DEPOSITION OF ANATASE TIO$_2$ PARTICLES AND THEIR CHANGE IN SURFACE AREA

Yoshitake Masuda
National Institute of Advanced Industrial Science and Technology (AIST), 2266-98 Anagahora,
Shimoshidami, Moriyama-ku, Nagoya 463-8560, Japan
* Corresponding Author: Y. Masuda, masuda-y@aist.go.jp

ABSTRACT

TiO$_2$ particles were synthesized in aqueous solutions at 50 °C for 30 min. They were assemblies of nano TiO$_2$ crystals. Transmission electron microscope revealed that nano-crystals grew along c-axis to form acicular shape. They showed high BET specific surface area of 270 m^2/g. For comparison, the particles were kept in the solution to progress further crystallization at 25 °C for 1 day. Crystal growth affected morphology, particle size and surface condition of TiO$_2$. Surface area was decreased to 168 m^2/g. Additionally, crystallite size increased with crystallization in the solutions. Crystallite size perpendicular to the (101), (004) or (200) planes increased 1.77 times, 2.94 times or 2.14 times, respectively. Especially, crystallite size perpendicular to (004) planes much increased compared to that of (101) or (200) planes. The solutions with low ion concentrations were suitable for anisotropic crystal growth along c-axis.

INTRODUCTION

TiO$_2$ particles have been applied for catalysts[1], photocatalysts[2-4], gas sensors[5,6], lithium batteries[7-9], biomolecule sensors[10] and dye sensitized solar cells[11,12]. They are prepared with flame synthesis[13,14], ultrasonic irradiation[15,16], chemical vapor synthesis[17], sol-gel methods[2,18-21], etc. However, they aggregate in high temperature annealing process. Surface area is drastically decreased by the aggregation. The particles obtained from sol-gel are covered with hydroxylated surfaces. They adsorb organic molecules unless extreme heat treatments or chemical dehydroxylation reactions.

Recently, liquid phase deposition of anatase TiO$_2$ was developed. For instance, ammonium hexafluorotitanate, boric acid and HCl were dissolved in water at 50 °C and kept for 12 h. Self-assembled monolayers(SAMs) were immersed in the solution to deposit anatase TiO$_2$. Deposition mechanism of TiO$_2$ films was evaluated in details[22].

In this study, anatase TiO$_2$ particles were synthesized in aqueous solutions at 50°C for 30 min with liquid phase crystal deposition method[23]. They were assembly of acicular anatase crystals. Each acicular crystal was grown along c-axis. The particles had nano-sized pores in the body. They showed high specific surface area of 270 m^2/g. For comparison, the particles were kept in the solution to progress further crystallization[24]. It affected on morphology, particle size and crystallite size. Specific surface area decreased to 168 m^2/g.

EXPERIMENT

Liquid phase crystal deposition of anatase TiO$_2$

Ammonium hexafluorotitanate ($[NH_4]_2TiF_6$) (Morita Chemical Industries Co., Ltd., FW: 197.95, purity 96.0%, 12.372 g) and boric acid (H_3BO_3) (Kishida Chemical Co., Ltd., FW: 61.83, purity 99.5%, 11.1852 g) were separately dissolved in deionized water (600 mL) at 50°C. Boric acid solution was added to ammonium hexafluorotitanate solution at a concentration of 0.15 and 0.05 M, respectively. The solution was kept at 50°C for 30 min using a water bath with no stirring.

TiO$_2$ particles 1: The solutions were centrifuged at 4000 rpm for 10 min (Model 8920, Kubota Corp.). Preparation for centrifugation, centrifugation at 4000 rpm, deceleration from 4000 to 0 rpm and preparation for removal of supernatant solutions took 4 min, 10 min, 10 min and 8 min, respectively. The particles contacted with the solutions, the temperature of which was gradually lowered for 32 min after maintaining 50°C for 30 min. Precipitates were dried at 60°C for 12 h after removal of supernatant solutions.

TiO$_2$ particles 2: For comparison, the solutions were then removed from water bath and left to cool for 1 day. Supernatant solutions were removed from the solutions. Heated water of 50 °C was added and mixed lightly. Heated water was used to dissolve residual ammonium hexafluorotitanate and boric acid. The solutions were then kept for 3 h to precipitate particles. Removal of supernatant solutions after dispersion in heated water of 50 °C was repeated three times. The solutions containing particles were dried on glass vessels at 60 °C for 1 day.

CHARACTERIZATION

The crystal phase of the particles was evaluated with X-ray diffractometer (XRD; RINT-2100V, Rigaku) with CuKα radiation (40 kV, 30 mA). Diffraction patterns were evaluated using JCPDS, ICSD (Inorganic Crystal Structure Database) data (FIZ Karlsruhe, Germany and NIST, USA) and FindIt. Morphology of TiO$_2$ was observed with transmission electron microscopy (TEM; JEM2010, 200 kV, JEOL) and a field emission scanning electron microscope (FE-SEM; JSM-6335F, JEOL Ltd.). Zeta potential and particle size distribution were measured with electrophoretic light-scattering spectrophotometer (ELS-Z2, Otsuka Electronics Co., Ltd.) with automatic pH titrator. Samples of 0.01 g were dispersed in distilled water (100 g) and ultrasonicated for 30 min prior to measurement. The pH of colloidal solutions was controlled by the addition of HCl (0.1 M) or NaOH (0.1 M). Zeta potential and particle size distribution were evaluated at 25°C and integrated 5 and 70 times, respectively.

Nitrogen adsorption-desorption isotherms were obtained using Autosorb-1 (Quantachrome Instruments) and samples were outgassed at 110°C under 10^{-2} mmHg for 6 h prior to measurement. Specific surface area was calculated with BET (Brunauer-Emmett-Teller) method using adsorption isotherms. Pore size distribution was calculated with BJH (Barrett-Joyner-Halenda) method using adsorption isotherms because an artificial peak was observed from BJH size distribution calculated from desorption branches. Pore size distribution was further calculated with DFT/Monte-Carlo method (N$_2$ at 77 K on silica (cylinder/sphere, pore, NLDFT ads. model), adsorbent: oxygen) using adsorption branches.

RESULTS AND DISCUSSION

(1) Morphology of TiO₂ particles

TiO₂ particles 1: The particles were consisted of nano-sized TiO_2 crystals (Fig. 1a). They were 100–200 nm in diameter. Particle surfaces had relief structures. They had open pores in their body. Nanocrystals had acicular shapes of about 5–10 nm in length. The FFT image analyses indicated that they were single phase of anatase TiO_2. The longer direction of the crystals was parallel to c-axis of anatase TiO_2. Crystal growth of anatase TiO_2 along the c-axis was previously observed in TiO_2 films[22]. Anisotropic crystal growth is one of the features of liquid phase crystal deposition. Crystallization of TiO_2 was effectively utilized to form assemblies of acicular nanocrystals in the process. Open pores and surface relief structures were successfully formed on the particles.

TiO₂ particles 2: The particles were about 800 nm in diameter (Fig. 1b). They were larger than that of above TiO_2 particles 1. TiO_2 particles 2 were kept in the solutions for longer period compared with TiO_2 particles 1. Crystallization of TiO_2 and aggregation of particles increased particle size. They were constructed of nano TiO_2 crystals to have nano-sized relief structure on surfaces. They were about 20-50 nm in diameter. The size was larger than crystallite size perpendicular to (101), (004) or (200) planes. Nanocrystals observed on surface of the particles would not be large single crystals but polycrystals constructed of several single crystals. The particles had nano-sized open pores on surfaces and insides surrounded by nano TiO_2 crystals. Nano-size open pores would contribute to high specific surface area and high adsorption properties which are required for photocatalysts, cosmetics, solar cells or sensors.

(2) Crystal phase of TiO₂ particles

TiO₂ particles 1: The particles showed X-ray diffraction peaks at 2θ = 25.1, 37.9, 47.6, 54.2, 62.4, 69.3, 75.1, 82.5 and 94.0° (Fig. 2a). They were assigned to the 101, 004, 200, 105 + 211, 204, 116 + 220, 215, 303 + 224 + 312 and 305 + 321 diffraction peaks of anatase TiO_2 (JCPSD No. 21-1272, ICSD No. 9852). The 004 diffraction intensity of randomly oriented particles is usually 0.2 times the 101 diffraction intensity as shown in JCPDS data (No. 21-1272). However, the 004 diffraction intensity of the particles deposited in our process was 0.36 times the 101 diffraction intensity. Additionally, the integral intensity of the 004 diffraction was 0.18 times the 101 diffraction intensity. Crystallite size perpendicular to the (101), (004) or (200) planes was estimated from the full-width half-maximum of the 101 or 004 peak to be 3.9 nm, 6.3 nm or 4.9 nm respectively. Elongation of crystals in the c-axis direction was also suggested by the difference in crystallite size.

TiO₂ particles 2: X-ray diffraction peaks were observed at 2θ = 25.3, 37.8, 47.9, 53.9, 62.5, 68.9, 69.8, 74.85, 82.3 and 94.4° (Fig. 2b). They were assigned to 101, 004, 200, 105 + 211, 204, 116, 220, 215, 303 + 224 + 312 and 305 + 321 diffraction peaks of anatase TiO_2. The 004 diffraction intensity was 0.61 times of 101 diffraction intensity. Integral intensity of 004 diffraction was 0.21 times of 101 diffraction intensity. Crystallite size perpendicular to (101), (004) or (200) planes were estimated from the full-width half-maximum of 101, 004 or 200 peak to be 6.9 nm, 18.5 nm or 10.5 nm, respectively. Elongation of crystals in the c-axis direction was suggested by the difference of crystallite size.

It is notable that crystallite size increased with crystallization in the solutions. Crystallite size perpendicular to the (101), (004) or (200) planes increased 1.77 times, 2.94 times or 2.14 times, respectively. TiO_2 particles 2 were kept in the solutions for longer period compared with TiO_2 particles 1. They showed that crystal growth proceeded in the solutions at room temperature for 1 day. Especially, crystallite size perpendicular to (004) planes much increased compared to that of (101) or (200) planes. Ion concentration of the solution decreased with time. TiO_2 particles 2 were kept in the solutions with low ion concentration. The solutions with low ion concentrations were suitable for anisotropic crystal growth along c-axis.

(3) N_2 adsorption characteristics of TiO_2 particles

TiO_2 particles 1: The desorption isotherm differed from adsorption isotherm in the relative pressure (P/P_0) range from 0.4 to 0.7, showing mesopores in the particles (Fig. 3a). BET specific surface area of the particles was 270 m^2/g. This is higher than that of commercial TiO_2 nanoparticles such as Aeroxide P25 (BET 50 m^2/g, 21 nm in diameter, anatase 80% + rutile 20%, Degussa), Aeroxide P90 (BET 90–100 m^2/g, 14 nm in diameter, anatase 90% + rutile 10%, Degussa), MT-01 (BET 60 m^2/g, 10 nm in diameter, rutile, Tayca Corp.) and Altair TiNano (BET 50 m^2/g, 30–50 nm in diameter, Altair Nanotechnologies Inc.)[25]. A high BET specific surface area cannot be obtained from particles having a smooth surface even if the particle size is less than 100 nm. A high BET specific surface area would be realized by the unique morphology of TiO_2 particles constructed of nanocrystal assemblies. Pore size distribution was calculated with the BJH method using adsorption isotherms. It showed a pore size distribution curve having a peak at ~2.8 nm and pores larger than 10 nm. TiO_2 particles would have mesopores of ~2.8 nm surrounded by nanocrystals. Pores larger than 10 nm are considered to be interparticle spaces. The pore size distribution also suggested the existence of micropores smaller than 1 nm. Pore size distribution was further calculated with the DFT/Monte-Carlo method. The model was in fair agreement with adsorption isotherms. Pore size distribution showed a peak at ~3.6 nm that indicated the existence of mesopores of ~3.6 nm. The pore size calculated with the DFT/Monte-Carlo method was slightly larger than that calculated from the BJH method because the latter method is considered to have produced an underestimation[26-28]. The pore size distribution also suggested the existence of micropores of ~1 nm, probably resulting from microspaces surrounded by nanocrystals and the uneven surface structure of nanocrystals. The particles were shown to have a large surface area as well as micropores of ~1 nm, mesopores of ~2.8–3.6 nm and pores larger than 10 nm, with N_2 adsorption characteristics. Assembly of acicular nanocrystals resulted in unique features and high surface area.

TiO_2 particles 2: Desorption isotherm was similar to adsorption isotherm showing absence or small amount of mesopores (2 nm ~ 50 nm) in the particles (Fig. 3b). BET specific surface area of the particles was estimated to 168 m^2/g from adsorption isotherm in the relative pressure (P/P_0) range from 0.02 to 0.07. The data followed a straight line of $F(x) = 20.7 x + 0.00741$. BET specific surface area was estimated from isotherm under $P/P0 = 0.1$ because isotherm of type I indicated micropores and y-intercept calculated from BET specific surface area above $P/P0 = 0.1$ was negative. BET specific surface area was higher than that of commercial TiO_2 nanoparticles. High surface area was realized by the unique morphology of the particles. Nano crystals in the particles would generate nano sized open pores and increase surface area. Additionally, surfaces of the particles were not covered with organic molecules or organic solvents because they were not included in the solutions.

The particles had surfaces of anatase crystals which are required to realize high performance as photocatalysts, cosmetics, solar cells or sensors.

CONCLUSION

TiO_2 particles were prepared at 50°C for 30 min using crystallization of anatse TiO_2 in aqueous solutions. They were assemblies of nanocrystals 5–10 nm and 100–200 nm in diameter. Nanocrystals grew anisotropically along the c-axis to form acicular shapes. BET specific surface area of the particles reached to 270 m^2/g. Pores of 1-4 nm were existed in the particles. It contributed high surface area. For comparison, the particles were kept in the solutions at room temperature for 1 day. TiO_2 particles further grew in the solutions. They were 800 nm in diameter and constructed of nano sized crystals. BET specific surface area was decreased to 168 m^2/g. Crystal growth affected morphology, particle size and surface condition of TiO_2. Additionally, crystallite size increased with crystallization in the solutions. Crystallite size perpendicular to the (101), (004) or (200) planes increased 1.77 times, 2.94 times or 2.14 times, respectively. Especially, crystallite size perpendicular to (004) planes much increased compared to that of (101) or (200) planes. The solutions with low ion concentrations were suitable for anisotropic crystal growth along c-axis.

REFERENCES

[1] T. Carlson and G. L. Giffin, "Photooxidation of Methanol using V2O5/TiO2 and MoO3/TiO2 Surface Oxide Monolayer Catalysts," J. Phys. Chem., 90(22), 5896-900 (1986).
[2] Z. B. Zhang, C. C. Wang, R. Zakaria, and J. Y. Ying, "Role of particle size in nanocrystalline TiO2-based photocatalysts," J. Phys. Chem. B, 102(52), 10871-78 (1998).
[3] R. Wang, K. Hashimoto, and A. Fujishima, "Light-induced amphiphilic surfaces," Nature, 388(6641), 431-32 (1997).
[4] W. Y. Choi, A. Termin, and M. R. Hoffmann, "The Role of Metal-Ion Dopants in Quantum-Sized TiO2 - Correlation between Photoreactivity and Charge-Carrier Recombination Dynamics," J. Phys. Chem., 98(51), 13669-79 (1994).
[5] N. Kumazawa, M. R. Islam, and M. Takeuchi, "Photoresponse of a titanium dioxide chemical sensor," J. Electroanal. Chem., 472(2), 137-41 (1999).
[6] M. Ferroni, M. C. Carotta, V. Guidi, G. Martinelli, F. Ronconi, M. Sacerdoti, and E. Traversa, "Preparation and characterization of nanosized titania sensing film," Sens. Actuators B: Chem., 77(1-2), 163-66 (2001).
[7] M. Wagemaker, A. P. M. Kentgens, and F. M. Mulder, "Equilibrium lithium transport between nanocrystalline phases in intercalated TiO2 anatase," Nature, 418(6896), 397-99 (2002).
[8] A. S. Arico, P. Bruce, B. Scrosati, J. M. Tarascon, and W. Van Schalkwijk, "Nanostructured materials for advanced energy conversion and storage devices," Nature Mater., 4(5), 366-77 (2005).
[9] Y. G. Guo, Y. S. Hu, and J. Maier, "Synthesis of hierarchically mesoporous anatase spheres and their application in lithium batteries," Chem. Commun., 26, 2783-85 (2006).
[10] H. Tokudome, Y. Yamada, S. Sonezaki, H. Ishikawa, M. Bekki, K. Kanehira, and M.

Miyauchi, "Photoelectrochemical deoxyribonucleic acid sensing on a nanostructured TiO2 electrode," Appl. Phys. Lett., 87(21), 213901-03 (2005).

[11] M. K. Nazeeruddin, F. De Angelis, S. Fantacci, A. Selloni, G. Viscardi, P. Liska, S. Ito, B. Takeru, and M. G. Gratzel, "Combined experimental and DFT-TDDFT computational study of photoelectrochemical cell ruthenium sensitizers," J. Am. Chem. Soc., 127(48), 16835-47 (2005).

[12] P. Wang, S. M. Zakeeruddin, J. E. Moser, R. Humphry-Baker, P. Comte, V. Aranyos, A. Hagfeldt, M. K. Nazeeruddin, and M. Gratzel, "Stable new sensitizer with improved light harvesting for nanocrystalline dye-sensitized solar cells," Adv. Mater., 16(20), 1806-11 (2004).

[13] P. W. Morrison, R. Raghavan, A. J. Timpone, C. P. Artelt, and S. E. Pratsinis, "In situ Fourier transform infrared characterization of the effect of electrical fields on the flame synthesis of TiO2 particles," Chem. Mater., 9(12), 2702-08 (1997).

[14] G. X. Yang, H. R. Zhuang, and P. Biswas, "Characterization and sinterability of nanophase titania particles processed in flame reactors," Nanostruct. Mater., 7(6), 675-89 (1996).

[15] J. C. Yu, J. G. Yu, W. K. Ho, and L. Z. Zhang, "Preparation of highly photocatalytic active nano-sized TiO2 particles via ultrasonic irradiation," Chem. Commun., 19, 1942-43 (2001).

[16] W. P. Huang, X. H. Tang, Y. Q. Wang, Y. Koltypin, and A. Gedanken, "Selective synthesis of anatase and rutile via ultrasound irradiation," Chem. Commun., 15, 1415-16 (2000).

[17] S. Seifried, M. Winterer, and H. Hahn, "Nanocrystalline titania films and particles by chemical vapor synthesis," Chem. Vap. Deposition, 6(5), 239-44 (2000).

[18] E. Scolan and C. Sanchez, "Synthesis and characterization of surface-protected nanocrystalline titania particles," Chem. Mater., 10(10), 3217-23 (1998).

[19] C. C. Wang and J. Y. Ying, "Sol-gel synthesis and hydrothermal processing of anatase and rutile titania nanocrystals," Chem. Mater., 11(11), 3113-20 (1999).

[20] S. D. Burnside, V. Shklover, C. Barbe, P. Comte, F. Arendse, K. Brooks, and M. Gratzel, "Self-organization of TiO2 nanoparticles in thin films," Chem. Mater., 10(9), 2419-25 (1998).

[21] H. Z. Zhang, M. Finnegan, and J. F. Banfield, "Preparing single-phase nanocrystalline anatase from amorphous titania with particle sizes tailored by temperature," Nano Lett., 1(2), 81-85 (2001).

[22] Y. Masuda, T. Sugiyama, W. S. Seo, and K. Koumoto, "Deposition Mechanism of Anatase TiO2 on Self-Assembled Monolayers from an Aqueous Solution," Chem. Mater., 15(12), 2469-76 (2003).

[23] Y. Masuda and K. Kato, "Nanocrystal Assembled TiO2 Particles Prepared from Aqueous Solution," Cryst. Growth Des., 8(9), 3213-18 (2008).

[24] Y. Masuda and K. Kato, "Aqueous Solution Synthesis of Anatase TiO2 Particles," J. Jpn. Soc. Powder Powder Metallurgy, 54(12), 824-27 (2007).

[25] R. K. Wahi, Y. P. Liu, J. C. Falkner, and V. L. Colvin, "Solvothermal synthesis and characterization of anatase TiO2 nanocrystals with ultrahigh surface area," J. Colloid

Interface Sci., 302(2), 530-36 (2006).

[26] M. Kruk and M. Jaroniec, "Gas adsorption characterization of ordered organic-inorganic nanocomposite materials," Chem. Mater., 13(10), 3169-83 (2001).

[27] P. I. Ravikovitch, S. C. Odomhnaill, A. V. Neimark, F. Schuth, and K. K. Unger, "Capillary hysteresis in nanopores: Theoretical and experimental studies of nitrogen adsorption on MCM-41," Langmuir, 11(12), 4765-72 (1995).

[28] C. Lastoskie, K. E. Gubbins, and N. Quirke, "Pore-Size Distribution Analysis of Microporous Carbons - a Density-Functional Theory Approach," J. Phys. Chem., 97(18), 4786-96 (1993).

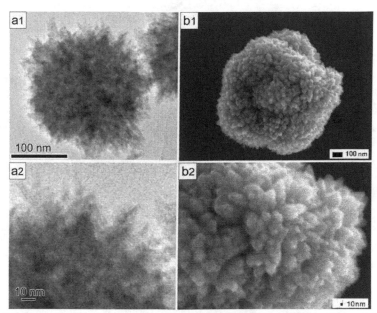

Fig. 1   Microstructures of anatase TiO$_2$ particles. (a) TEM images of TiO$_2$ particles 1 that were synthesized at 50°C for 30 min. (b) SEM images of TiO$_2$ particles 2 that were synthesized at 50°C for 30 min and at 25°C for 1 day.

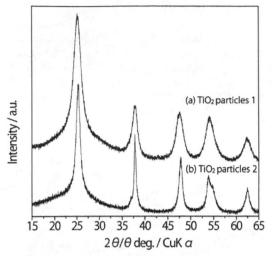

Fig. 2   XRD diffraction pattern of anatase TiO$_2$ particles. (a) TiO$_2$ particles 1 that were synthesized at 50°C for 30 min. (b) TiO$_2$ particles 2 that were synthesized at 50°C for 30 min and at 25°C for 1 day.

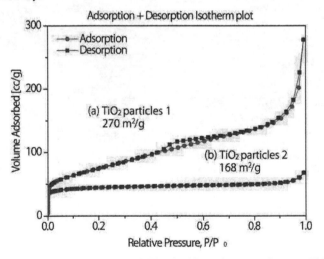

Fig. 3   N$_2$ adsorption-desorption isotherm and BET specific surface area of anatase TiO$_2$ particles. (a) TiO$_2$ particles 1 that were synthesized at 50°C for 30 min. (b) TiO$_2$ particles 2 that were synthesized at 50°C for 30 min and at 25°C for 1 day.

SINGLE STEP SYNTHESIS AND SELF-ASSEMBLY OF MAGNETITE NANOPARTICLES

M. Hoffmann, R. von Hagen, H. Shen, S. Mathur*

Institute of Inorganic Chemistry
University of Cologne
D-50939 Cologne
Germany
*sanjay.mathur@uni-koeln.de

ABSTRACT

Mono-disperse magnetite nanoparticles (d ~ 6.9 nm, σ = 2.9 %) were synthesized by the thermal decomposition of iron(III) oxalate in a high boiling solvent followed by the addition of oleic acid as a capping agent. Mechanistic aspects of the decomposition of the precursor were investigated by FT-IR spectral studies of aliquots taken at different synthesis temperatures. SEM, TEM and HRTEM analyses revealed the tendency of the magnetite particles to aggregate into larger crystals. The dispersability of magnetite nanoparticles in water was achieved by exchanging the oleic ligands with tartaric acid. Due to the ligand-ligand interactions, superstructures are constructed based on the self-assembly; this was confirmed by the SEM-cross-sectional analysis of the supra-crystals using the FIB method. The dispersion of magnetite nanoparticles in water allowed their use as colloidal ink for ink-jetted fabrication of micro/nano device structures. The I-V curves of the "bridge" of magnetite particles printed between two Ag electrodes showed ohmic contacts. Such printable nanostructures possess potential for gas sensing application.

INTRODUCTION

Due to their finite crystal size and the high surface to volume ratio, iron oxide nanomaterials possess unique magnetic and electrical properties.[1-5] Magnetite (Fe_3O_4) is ferromagnetic and shows half-metallic behavior due to electron hopping between Fe^{2+} and Fe^{3+} sites.[6, 7] Since the magnetic properties of this material strongly depend on the particle size, lattice defects and Fe:O stoichiometry, the controlled preparation of nano-scaled Fe_3O_4 offers promising potential towards the development of advanced magnetic material as well as in their 2D or 3D arrays with distinct properties which can be used for devices such as LED's[8], or magnetic storage media[9, 10]. Further, the inter-particle interaction in the densely packed neighbors can change the properties, when compared to dipersed particles, however the driving forces of self-assembly are not well understood and controlling the processes will be a challenging task[11]. Here we report a novel synthetic approach of monodisperse magentite nanoparticles and results related to assembling and transport behaviors in surface modified Fe_3O_4 NPs.

EXPERIMENTAL

Iron(III) oxalate was purchased from Sigma-Aldrich . All experiments were performed in a modified Schlenk type vacuum assembly, taking stringent precautions against atmospheric moisture. All solvents used for the synthesis were purified and dried by standard methods and stored over sodium. The synthesis of Fe_3O_4 NPs was performed using the Heating-Up method.[12] In a typical synthesis, 242 mg (0.5 mmol) iron(III) oxalate was stirred in 15 ml 1-octadecene for 5 minutes, then 314 μL (1.0 mmol) oleic acid, 194 μL (0.5 mmol) oleylamin and 75 mg (1.00 mmol) trimethylamine N-oxide (TMAO) were added at room temperature and the reaction mixture was gradually heated-up in the temperature range of 300°C-320°C. After 1 h of heat treatment under argon flux, the solution was cooled down to room temperature and 30 ml ethanol were added. The iron oxide particles were obtained as a black sediment, which could be removed through centrifugation (4400 rpm, 20 min). For the ligand-exchange reaction, 20 mg oleic acid capped particles were dissolved in 1 mL toluene and stirred over night with 20 mg tartaric acid dissolved in 1 ml methanol. The particles were collected by centrifugation (4400 rpm, 2 min), dissolved in 5 ml water and freeze-dried for storage. Scanning electron microscopy (SEM) and focused ion beam (FIB) analyses were performed on a Zeiss Neon 40 Cross Beam system. The suspension of magnetite nanoparticles were sonicated and deposited on carbon coated copper grids and measured on a Phillips CM 300 transmission electron microscope. To study the I-V characteristics, a linear Fe_3O_4 "bridge" was printed on the Ag electrodes using an ink-jet printing system (Microdrop MD-P-802) and tested towards the gas sensing properties on a PC-controlled gas sensing characterization unit. The temperature-dependent I-V curves were recorded on a Keithley 2400 source meter unit.

RESULTS AND DISCUSSION

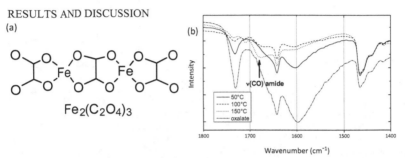

Figure 1. (a) Postulated structure of iron(III) oxalate with two terminal and one bridging oxalate groups. (b) FT-IR spectra of pure iron(III) oxalate and aliquots taken from the reaction mixture at different temperatures during the heating-up process.

The thermal decomposition of iron carboxylates is known as an efficient method for the synthesis of mono-disperse iron oxide nanostructures.[12-17] Here we demonstrate the synthesis of magnetite nanoparticles using an iron(III)-oxalate hexahydrate precursor using the Heating-Up method, followed by an in-situ surface modification, which suppresses the agglomeration effects. Although the single crystal XRD analysis was not fruitful, due to the instability of the crystals, a dimer bridged by a tetradental oxalate anion is proposed on the basis of spectral data. Infrared and Raman investigations support the presence of terminal oxalate groups (Fig. 1a). Based on NMR studies, oleylamine was found to have an activating effect on the decomposition process of metal carboxylates by nucleophilic attack and elimination of the amide leaving metal oxides and hydroxides as the possible products, which form the real precursor to iron oxide and might be decomposed with a different mechanism, when compared to pure iron(III) oxalate. This leads to the reduction of the decomposition temperature of about 10°C and allows the processing at lower temperatures. The formation of the amide by elimination of the carboxyl group could be proven by the appearance of the typical amide $v(CO)$ IR-band at 1675cm^{-1} starting from a reaction temperature of 100 °C (Fig. 1b). The concomittant reduction of two $v(CO)$ bands (1730 cm^{-1}, 1600 cm^{-1}) of the oxalate underline the suggested elimination mechanism. Further efforts are necessary for the better understanding of the decomposition pathways that may provide insight into the free-energy landscape of precursor to oxide conversion, and could lead to a better control over the nucleation and growth processes.

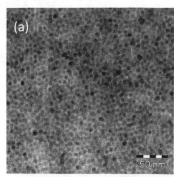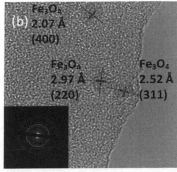

Figure 2. (a) TEM analysis confirmed the narrow size distribution of Fe_3O_4 NPs; (b) HR-TEM and electron diffraction images of Fe_3O_4 NPs confirm the crystallinity of Fe_3O_4 NPs.

TEM observations (Fig. 2) revealed a dense packing of oleic acid capped magnetite particles with a diameter of 6.9 nm and exhibiting a narrow size-distribution of 2.9 %. The high crystallinity and phase identification of magnetite NPs were confirmed by HR-TEM and electron diffraction analyses showing the lattice fringes and electron diffraction rings ((220), (311), (400)) corresponding to Fe_3O_4 NPs. The dense packing of the magnetite nanocrystals can be induced possibly by entropic, *van der Waals* forces and/or dipole-dipole (U_{dd}) interactions between the magnetic domains. The dipolar magnetic force is weak for spontaneous agglomeration in diluted solutions (< 1 kT), but it can play an important role at higher concentrations during solvent evaporation ($U_{dd} \sim r^{-3}$)[11].

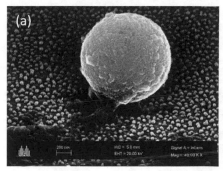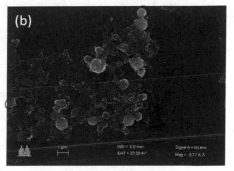

Figure 3. SEM images of (a) an assembled Fe_3O_4 super-sphere and (b) a typical distribution after evaporation of a diluted solution.

The as-prepared magnetite NPs were modified by tartaric acid in order to study the self-assembly behavior. The SEM images (Fig. 3) of tartaric acid capped particles showed the presence of large spherical structures with size ranging from 200 nm to 1 μm. It was found that these super-spheres were also formed in the diluted solutions in which the magnetic dipole-dipole interactions should be weak (Figure 3). Hence the driving force of formation must be highly influenced by the ligand-ligand interactions of tartaric acid ligands, as the spheres were not observed in the solutions of hydrophobic (oleate-coated) nanoparticle solutions under similar time periods. The formation of ball-like structures is faster in higher concentrations due to higher collision probabilities and resulting adsorbtion flux that is terminated by the thermodynamic balance between the ripening process leading to spherical suprastructures and the more diluted solution.

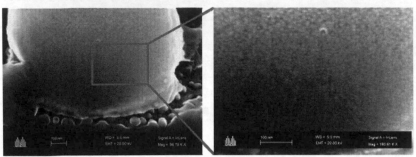

Figure 4. SEM images of cross-sectional surface of Fe_3O_4 super-spheres prepared by focused ion beam (FIB) show the dense packing of single particles in the super structure.

Magnetite particles with sizes below 10 nm are known to be superparamagnetic. However densely packed spheres composed of assembled magnetic nanoparticles showed collective magnetic properties due to dipolar interactions of 3D ordered superparamagnetic domains.[18, 19] A cross sectional analysis of a magnetite supra-structure by focused ion beam (FIB) showed that the assembled balls have a continuous dense packing of nanoparticles and do not form hollow structures (Fig. 4). This observation implies that collectic magnetic properties could be present and affect a change of the hysteresis behavior.

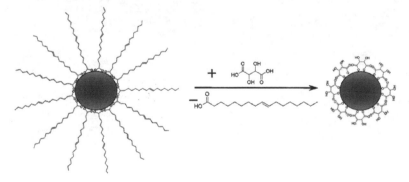

Figure 5. Schematic illustration of the Fe_3O_4 NP surface modification by exchange of the oleic acid ligands with tartaric acid to obtain water-based dispersions.

Water dispersible particles are used as nano-inks to print functional patterns (circuits) for nanodevice applications, which requires high surface tension disperse media for clean drop shapes and good stability of the solution to avoid closure of the ink-jet nozzles. For this reason the C_{18} groups of the apolar oleic acid surfactants were removed by ligand exchange with polar tartaric acid groups through mixing the particles with tartaric acid in a biphasic system (toluene/methanol) (Fig. 5). The lower pk_s of tartaric acid ($pk_1 = 3.03$, $pk_2 = 4.36$)[20] in comparison to oleic acid ($pk_s = 4.70$)[21] causes its full deprotonation to the tartaric dianion and protonation to oleic acid. This excess of available surfactants moves the equilibrium towards the hydrophilic capped particles. These can form hydrogen bonds with the polar solvent (e.g. water) thereby stabilizing the particles in solution. Aqueous dispersions of modified crystals were stable for one day at lower concentrations (0.5 mg/1 ml) whereas redispersion in organic solvents, such as dichloromethane, was not successful. Solutions with higher particle concentrations (> 1 mg/ml) were stable up to three hours before gradual flocculation began (self-assembly of spheres). The dispersability of the magnetite particles in water enabled the application as colloidal ink and thus a flexible and fast patterning of microstructures with a resolution of 20-100 μm, demonstrated on various substrate materials. The linear magnetite structures could be formed by printing magnetite ink on silver electrodes (Fig. 6).

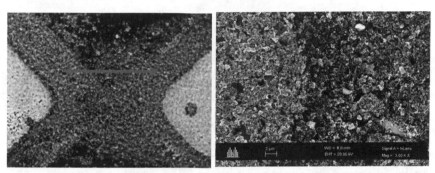

Figure 6. Optical (printing direction is illustrated by an arrow) and SEM images of ink-jet printed conductive and porous Fe$_3$O$_4$ linear structures on two silver electrodes.

For proper contacts between particles, the inter-particle separation has to be small whereby close packing of monodisperse particles is beneficial. This leads to a weak conductivity of oleic acid capped particles with long and nonconductive hydrocarbon chains (C$_{18}$). In this case the organic surfactant shell has got a thickness of about 1 nm and acted as non-conductive spacer.[22] The charge transport is limited by the energetic barrier of electron tunneling processes between single crystal domains. The resistance increases with larger inter-particle distances and can lead to insulating behavior.[23, 24] The chain length of tartaric acid is significantly shorter (C$_4$) which should lead to thinner ligand shells. Additionally, the previously observed strong ligand-ligand interactions of the particles via hydrogen bonding of hydroxyl groups led to dense packing of the particle assemblies and formed super-spheres (Fig 3).

Figure 7 shows a linear behavior of the printable assembly of magnetite particles, which behaves as an ohmic resistor and the resistance at room temperature could be estimated to be 4.5 kΩ. Therefore, the conductivity is about one order of magnitude lower than that found in oleic acid capped magnetite particles (10^2 Ω), annealed at 400°C-650°C. Apparently, the insulating surfactant was thermally desorbed/decomposed leaving inter-particle voids and disrupting the dense packing, which can pose a higher tunneling barrier as the resistance is determined by the electron hopping barrier between Fe^{2+} and Fe^{3+} sites due to the direct inter-particle contacts.[23, 25] This observation confirms that the interparticle distances between tartaric acid capped magnetite particles are small enough to enable the electron hopping process and thus enables the flexible patterning of conductive magnetite material without any heat treatment. This room temperature process enables the use of thermally sensitive substrates like e.g. polycarbonate.

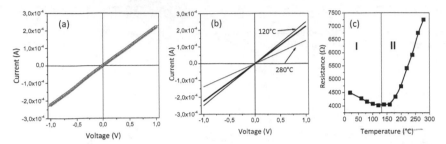

Figure 7. RT and variable temperature I-V curves (a and b) and resistance curves of Fe_3O_4 linear structures showed the ohmic behavior (c) of printed devices. A typical semiconductor behavior is observed up to 120°C followed by increasing resistivity at higher temperatures.

Based on the temperature-dependent I-V curves of the printed magnetite structures, the resistances was derived by the calculation of the different I-V slopes. It is evident that the thermal changes of resistances can be divided in two regions: (i) semiconducting behavior up to 120 °C and (ii) oxidation of Fe_3O_4 to Fe_2O_3 phase with higher resistivity.

These preliminary studies indicate the tremendous potential of surface modified magnetite NP as nano-ink for fast and flexible integration of magnetic and conductive materials via the microprinting technique. The porous structure (Fig. 6) combined with the ohmic behavior also denoted potential applications in the field of gas sensor.

CONCLUSION

In summary, mono-disperse mangetite nanoparticles have been synthesized by thermal decomposition of iron(III)-oxalate using the Heating-Up method. The decomposition pathway was monitored by FT-IR observations and the nanoparticles were modified with tartaric acid ligands to achieve dispersibility in water. Given the spherical nature and narrow size-dispersion, the particles showed a tendency of self-assembly in spherical suprastructures in solution without any external stimulus. The control of this effect could lead to dense packed superstructures with tunable collective magnetic and/or electrical properties. Additionally, the magnetite NPs modified with tartaric acid showed great potential in the prepartion of magnetic nano-ink. The I-V curves of printed magnetite structures showed good conductivity indicating their application potential, for instance in gas sensors.

ACKNOWLEDGEMENTS

Authors are thankful to the University of Cologne and the European Commission for the financial assistance within the framework programme FP7 for the projects 'S3' (EU-Russia Collaboration) and NANOMMUNE (EU-USA collaboration), as well as BMBF (MONOGAS). Thanks are also due to Dr. Belkoura (Department of Chemistry, University of Cologne) and Mr. Huehne (Uni. Bonn) for the TEM analyses.

REFERENCES

1. N.A. Frey, S. Peng, K. Cheng, S. Sun, *Chem. Soc. Rev.*, 2009, **38**, 2532.
2. C.B. Murray, C.R. Kagan, M. G. Bawendi, *Annu. Rev. Mater.*, 2000, **30**, 545.
3. N.R. Jana, Y. Chen, X. Peng, *Chem. Mater.*, 2004, **16**, 3931.
4. A.-H. Lu, W. Schmidt, N. Matoussevitch, B. H. Bönnermann,B. Spliethoff, B. Tesche, E. Bill, W. Kiefer, F. Schüth, *Angew. Chem.*, 2004, **116**, 4403.
5. Y.-W. Jun, Y.-M. Huh, J.-S. Choi, J.-H. Lee, H.-T. Song, S. Kim, S. Yoon, K.-S. Kim, J.-S. Shin, J.-S. Suh, J. Cheon, *J. Am Chem. Soc.*, 2005, **127**, 5732.
6. H. Zeng, C.T. Black, R.L. Sandstrom, P. M. Rice, C.B. Murray, S. Sun, *Phys. Rev B*, 2006, **73**, 020402(R).
7. W. Weiss, W. Ranke, *Progress in Surface Science*, 2002, **70**, 1.
8. S. Coe, W. Wook, M. Bawendi, V. Bulovic, *Nature*, 2002, **420**, 800.
9. S. Sun, C.B. Murray, D. Weller, L. Folks, A. Moser, *Science*, 2000, **287**, 1989.
10. D. E. Speliotis, *J. Appl. Phys*, 1999, **85**, 4325.
11. K.J.M. Bishop, C.E. Wilmer, S. Soh, B. A. Grzybowski, *Small*, 2009, **5**, 1600.
12. S.G. Kwon, Y. Piao, J. Park, S. Anagappane, Y. Jo, N.-M. Hwang, J.-G. Park, T. Hyeon, *J. Am Chem. Soc.*, 2007, **127**, 12571.
13. T.-D. Nguyen, T.-O. Do, *J. Phys. Chem. C*, 2009, **113**, 11204.
14. N. Bao, L. Shen, W. An, P. Padhan, C.H. Turner, A. Gupta, *Chem. Mater.*, 2009, **21**, 3458.
15. C.-J. Chen, H.-Y Lai, C.-C. Lin, J.-S. Wang, R.-K. Ciang, *Nanoscale Res. Lett.*, 2009, **4**, 1343.
16. L.M. Bronstein, X. Huang, J. Retrum, A. Schmucker, M. Pink, B.. D. Stein, B. Dragena, *Chem. Mater.*, 2007, **19**, 3624.
17. T. Hyeon, *Chem. Commun.*, 2003, **9**, 927.
18. V. Russier, C. Petit, M. P. Pileni, *J. Appl. Phys.*, 2003, **93**, 10001.
19. M. P. Pileni, *J. Phys D: Appl Phys.*, 2008, **41**, 134002.
20. J. Barbosa, J.L. Beltràn, V. Sanz-Nebot, *Analytica Chimica Acta*, 1994, **288**, 271.
21. M.B. Sankaram, P.J. Brophy, W. Jordi, D. Marsh, *Bioch. Bioph. Act.*, 1990, **1021**, 63.
22. S.I. Rybchenko, A.K.F. Dyab, S.K.Haywood, I.E. Itskevich, V. N. Paunov, *Nanotechnology*, 2009, **20**, 425607.
23. S. Jang, W. Kong, H. Zeng, *Phys. Rev. B*, 2007, **76**, 212403.
24. H. Zeng, C.T. Black, R.L. Sandstrom, P.M. Rice, C.B. Murray, S. Sun, *Phys. Rev. B*, 2006, **73**, 020402.
25. R. M. Cornell, U. Schwertmann, 'The iron oxides: structure, properties, reactions occurences and uses', WILEY-VCH, 2003.

PRINTABLE SILVER NANOSTRUCTURES: FABRICATION AND PLASMA-CHEMICAL MODIFICATION

R. von Hagen, M. Hoffmann, T. Lehnen, L. Xiao, D. Zopes, H. Shen and S. Mathur*

University of Cologne, Department of Chemistry
Institute of Inorganic and Materials Chemistry
Greinstraße 6
D-50939 Cologne
Germany

ABSTRACT

Printed nanostructures consisting of metallic silver particles were fabricated by printing and subsequent plasma-chemical reduction of appropriate silver salt. The method is applicable for temperature and chemically-sensitive polymer substrates as well, due to low processing temperatures and non-oxidizing nature of argon plasma. Detailed structural (XRD, SEM, FIB) and electrical studies on the silver nanostructures showed the resistivity of silver patterns (3.99×10^{-6} Ωcm) to be comparable with that of bulk silver (1.6×10^{-6} Ωcm). The printed silver patterns can be used for electronic circuitry used in applications such as gas sensors or for local patterning of metal catalysts useful for growing ZnO nanowires by chemical techniques as demonstrated in this work.

INTRODUCTION

Ink-jet printing is a versatile method for printing nanostructures by using numerous organic and inorganic inks and a powerful tool for rapid micropatterning on various kinds of substrates.[1] Subjected to the targeted applications the chemical composition of inks reaches from polymers, aqueous salt or precursor solutions to different dispersions of nano-sized metals or metal oxides.[1-4] Examples of ink-jet printing in the fabrication of functional microstructures include printed electronics,[1] lubricants for micromechanical parts[1] or defined polymer deposition for manufacturing multicoloured displays.[2] Additionally, ink-jet printing allows the fast, flexible and space-resolved integration of nanostructures in devices. Due to the bandwidth of their applications printed electronic circuits on flexible polymer foils are of increasing interest.[5] In the last few years, a lot of work focused on printing of metal-colloids or precursors to obtain conductive circuits after sintering has been reported.[1-3] The most simple approach is the heat-treatment for metallization and sintering processes. The development of new metal-complexes as precusors or stable colloidal solutions with high solid (metal) fractions and low sintering temperatures is of special interest. Apart from heat treatment[6] different sintering methods like laser,[7] microwave,[8] UV[9] and plasma[10] treatments have been described. Due to temperature sensitivity of polymer substrates, soft processing methods are currently required. Commercially available inks generally contain relatively expensive metal-colloids or metal-complexes of silver or copper. In 2005 Liu et al.[11] described the formation of silver tracks by thermal decomposition of a silver salt ($AgNO_3$), however the high processing temperature (~ 300 °C) required for the reduction ($Ag^+ \rightarrow Ag^0$) are not suitable for most of the transparent polymers. Plasma are widely used in semiconductor device fabrication e.g., for reactive ion etching (RIE), plasma enhanced chemical vapour deposition (PECVD), or sputtering. PECVD is extensively used for coating of substrates with various organic and inorganic functional layers.[12] Depending on the nature of gas species, plasma has different effects e.g., oxidizing[13] in case of O_2

as well as reducing in case of Ar atmosphere.[14] The use of $AgNO_3$ in printing electronic circuits is promising because it is easily available and highly soluble in water. Herein we present the synthesis and microstructures of silver patterns, formed by plasma-induced metallization (PIM) of $AgNO_3$. The formation mechanism and electrical resistivity of silver nanostructures as a function of plasma power are studied by cross-sectional SEM and four-probe measurements.

EXPERIMENT

In this study a water-based ink containing 50 wt.% of $AgNO_3$ (pure Ph. Eur., AppliChem, Germany) was used as silver source. Printing was carried out on a commercial printer (Model MD-P-802 of Microdrop Technologies, Germany) equipped with a dispensing-head possessing a nozzle (diameter, 50 µm) and a heating stage set to 120 °C. To investigate the formation of silver by X-Ray powder diffraction (XRD) analysis, a 50 x 50 square dot matrix with an internal drop-distance of 80 µm was printed on glass slides (VWR International, ECN 631-1578). The metallization of printed $AgNO_3$ patterns was examined by recording XRD on a STOE-STADI MP X-Ray diffractometer using CuK_α radiation. The diffraction patterns were simulated for the calculation of the crystallite size with STOE WinXPOW software. Dot and line patterns were printed on polyimide (PI) film (Kapton® 500 FPC, 127 µm, DuPont, USA) and glass slides (Marienfeld, Germany) and imaged by optical (Nikon Eclipse LV 150) and scanning electron microscopes (Zeiss Neon 40 Cross Beam). The later was also used for focused ion beam (FIB) assisted cross-sectional SEM measurements. The reduction of silver salt was carried out in a plasma reactor (Domino, Plasma Electronic, Germany), applying argon plasma with a constant gas flow of 20 sccm for 10 mins at different power. The conductivity of the silver tracks was determined by four-probe measurements using Keithley 2400 source meter (Keithley Instruments, USA), whereas their profile was measured using a chromatic sensor profiler (Fries Research & Technology, Germany). The ZnO nanocrystals were grown solvothermally on Si(100) substrates patterned with nanostructured silver as seed layer (an aqueous 0.01 M $Zn(NO_3)_2$ x 6 H_2O (reagent grade, Merck, Germany) solution in 50 ml sealed autoclave (Berghof, Germany).

RESULTS AND DISCUSSION

The water based $AgNO_3$ ink was light stable, easy to print in different dot and line patterns. Figure 1a shows a dot pattern of $AgNO_3$ on polyimide substrate before and after plasma treatment. The plasma-chemical modification of $AgNO_3$ structures resulted in the decomposition of metal salt to produce metallic particles (eq. 1).

$$AgNO_3 \xrightarrow{Ar^+} Ag^0 + NO_x \qquad (1)$$

To study the influence of plasma power on the formation of metallic silver, $AgNO_3$ structures on glass slides were treated in argon plasma for 10 mins using different current values (30-150 W). After the plasma processes the substrate temperature was monitored using an infrared thermometer (Voltcraft IR-360, Conrad Electronic SE, Germany), which showed only slight increase in substrate temperature (up to 55 °C) even at higher plasma energy (150 W) indicating that this reduction method is suitable for temperature sensitive polymer substrates.

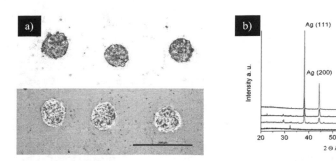

Figure 1. a) Printed dot pattern on polyimide substrate before (upper) and after (lower) plasma treatment (scale bar: 200 μm) and b) XRD patterns of AgNO₃ plasma-treated at various powers.

The XRD patterns (Fig. 1b) confirmed the chemical reduction process outlined in equation 1 and the subsequent metallization of AgNO₃ as a function of the applied plasma powers. The accelerated plasma ions are able to reduce the Ag⁺ and oxidize the NO₃⁻ to form elemental silver and release nitrogen oxides at very low temperatures whereby the exact reaction mechanism demands further investigations. In contrast to the oxidizing feeding gas like O_2 argon was chosen due to its inert nature in order not to induce any chemical reaction on the polymer substrate, and thus, to not affect the functional properties of the polymer film. With increasing plasma energy, the diffraction peaks of silver-nitrate disappeared and new peaks corresponding to elemental silver were obtained when plasma power of 150 W was applied for 10 mins. To study the influence of plasma power on the crystalline size (d_{hkl}) of the silver nanostructures, the Scherrer-formula (eq. 2) was used, where k is a constant (set to 0.9), λ the wavelength of CuK$_\alpha$ radiation, β the half width of the diffraction peak and θ the angle of the diffraction peak.

$$d_{hkl} = \frac{k\lambda}{\beta \cos\theta} \qquad (2)$$

The calculated crystallite sizes against the plasma powers are plotted in Figure 2. By increasing the power from 30 to 100 W the crystallite size showed only a slight increase. When the plasma power was set to 150 W a clear increase in the average crystallite sizes was observed e.g., the crystallite size (d_{111} direction) increases from 26 nm at 30 W to 61.96 nm at 150 W. In all cases the silver crystals show a texture along the d_{111} direction, apparently an influence of the crystal chemistry that favors anisotropic growth when particles crystallize out of a melt, whereby surfaces with lowest energy ((111) for e.g.) grow faster. Application of a higher plasma power may results in a higher acceleration of plasma ions and a deeper penetration of the silver nitrate layer and thereby larger crystallites will be created. As the (111) direction possesses the lowest energy of the crystallite facets, the growth of crystallites is elongated in this direction at all power rates.

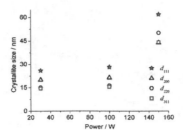

Figure 2. Influence of applied plasma powers on the crystallite size of the produced silver nanostructures.

To analyze the dependence of plasma power on electrical resistivity of silver nanostructures, tracks of silver-nitrate were printed on polyimide foil replicating a software programmed line pattern, fabricated with an internal drop distance of 40 μm. Figure 3a shows the profile of the silver track after treatment at 150 W for 10 mins where the height z is plotted versus the line width. The specific resistivity ρ was determined by a four-probe measurement (eq. 3), where R is the measured resistivity, s the cross sectional area and l the length of the line.

$$\rho = R \cdot \frac{s}{l} \tag{3}$$

The specific resistivity of the silver track decreases with increasing plasma powers from 2.21×10^{-3} Ωcm at 30 W to 4.39×10^{-5} Ωcm at 150 W (Fig. 3b), that accounts for 28 times of the resistivity of bulk silver. The reason for the different resistivity values at the different processes can be explained by the different microstructures and crystallite sizes. At 30 W the Ag nanostructures were composed of smaller crystallite sizes and a porous microstructure whereas at 150 W dense silver layers were obtained giving rise to higher conductivity.

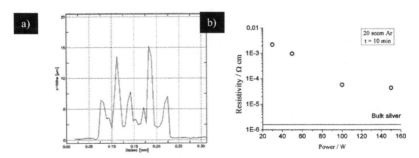

Figure 3. a) Profile of a printed line transferred to silver at 150 W. The average height is about 5.5 μm and the line width 170 μm. b) Specific resistivity plotted against the applied plasma power. The process time was set to 10 mins and the electrical resistivity of bulk silver (line) is given for comparison.

The comparatively high resistivity observed for printed and plasma transformed silver nanostructures can be due to the thickness of the printed silver lines and the microstructure. The cross-sectional SEM images of Ag lines metalized at 30 and 150 W showed that the argon plasma reacts chemically on the surface of the silver nitrate and the surface below is not completely decomposed (Fig. 4), validating the assumption that no thermal reduction occurs on the substrate. Due to the relative high thickness of the printed line patterns (~ 5.5 μm), the AgNO$_3$ was not metalized totally at 150 W. The silver layer formed at 150 W was thick enough (> 1 μm) and should be able to generate conductive pathways for charge carriers and to reduce scattering losses at the grain boundaries.

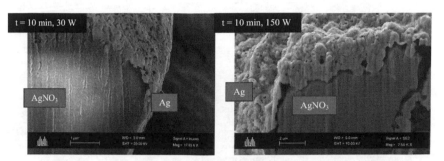

Figure 4. Formation of silver film on the surface of printed AgNO$_3$. The thickness can be regulated by varying the applied plasma power.

The porous microstructure observed in the silver tracks (Fig. 4) is possibly due to the effect of plasma ion bombardment and release of NO$_x$ gaseous species. At 30 W, tiny silver structures are directly formed on the surface of the AgNO$_3$, whereas at 150 W a small gap between the silver and silver nitrate film is generated. The silver layer formed at 30 W is highly porous whereas the layer formed at 150 W appears as dense silver foam. The average thickness of the silver layer (220 nm at 30 W and 500 nm at 150 W) was used to circulate the effective cross sectional area s^* of the printed silver lines, which further decreased the specific resistivity. This substitution was done for a plasma power of 30 and 150 W and the value decreases down to 3.99 x 10^{-6} Ωcm for the process with 150 W, which is only 2.5 times higher than that of bulk silver (Fig. 5) and suggests the high quality and metallic nature of our Ag patterns.

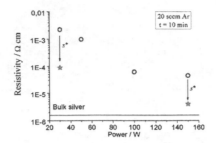

Figure 5. The corrected specific resistivity of the metallized samples as a function of the used plasma power.

The porous silver structures were further used as nucleation seeds to grow ZnO nanocrystals. Printing of thin line patterns on silicon wafers and their plasma metallization could be used for regio-selective hydrothermal growth of ZnO nanocrystals. Figure 6a shows a cross sectional SEM image of porous silver formed by metallization of $AgNO_3$ at 150 W on a glass substrate. The different nanostructures such as nano-bars on silicon and foam on polyimide (Fig. 4b) can be explained by thinner coating of $AgNO_3$ on silicon which is easily metallized and the better thermal conductivity of the glass substrate when compared to polyimide substrates.

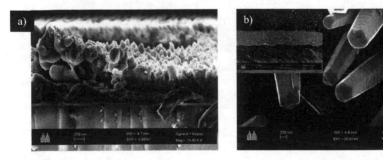

Figure 6. Cross sectional SEM images of a) porous silver formed by metallization of $AgNO_3$ in argon plasma and b) its use as nucleation seed for hydrothermal growth of ZnO on Si(100).

Figure 6b shows hydrothermally grown ZnO nanocrystals on a silver line patterned Si(100) substrate. The crystals are homogeneous, densely packed, around 200 nm in diameter and several micrometer in length. In 2007 Postels et al. reported the regio-selective growth of ZnO nanocrystals on Ag patterns produced by e-beam evaporation technique.[16] The fast and flexible ink-jet printing method enables the regio-selective growth of ZnO without using photolithographic or electron beam methods[15] and makes site-specific growth of nanostructures accessible.

CONCLUSIONS

Conductive silver patterns on the polyimide Kapton® 500 FPC were fabricated by ink-jet printing of silver nitrate and its metallization in argon plasma. Plasma chemical transformation of printable structures is suitable for almost all polymer substrates. The resistivity of achieved silver nanostructures (3.99×10^{-6} Ωcm) is only 2.5 times higher than that of bulk silver indicating the application potential of the combinational approach involving ink-jet printing and plasma-modification for developing conductive electrodes on polymers. Additionally, the method allows the regio-selective printing of metal patterns which were used as seed layers to grow ZnO nanocrystals on Si(100) and can be exploited for site-specific growth of nanostructures.

ACKNOWLEDGEMENT

Authors are thankful to the University of Cologne and the Federal Ministry of Education and Research (BMBF; KoLiWIn 55102006) for the financial support. We would like to thank DuPont High Performance Materials for providing the polyimide Kapton® 500 FPC. Thanks are due to Thomas Rügamer, Nurgül Tosun and Ramona Gerwig for the assistance with the plasma reactor and Alexander Voigt for profile measurements.

REFERENCES

[1]P. Calvert, *Chem. Mater.*, **13**, 3299 (2001).

[2]E. Tekin, P. J. Smith, U. S. Schubert, *Soft Matter*, **4**, 703 (2008).

[3]H.P Le, *J. Imaging Sci. Tech.*, **42**, 49 (1998).

[4]www.microdrop.de

[5]Market forecast, IDTechEx 2006.

[6]K.-S. Chou, K.-C. Huang and H.-H. Lee, *Nanotechnology*, **16**, 779 (2005).

[7](a) S. H. Ko, H. Pan, C. P. Grigoropoulos, C. K. Luscombe,J. M. J. Frechet and D. Poulikakos, *Appl. Phys. Lett.*, **90**, 141103 (2007). (b) K. C. Yung, S. P. Wu, H. Liem, *J. Mater. Sci.*, **44**, 154 (2009).

[8](a) J. Perelaer, B.-J. de Gans and U. S. Schubert, *Adv. Mater.*, **18**, 2101 (2006). (b) R. Roy, D. Agrawal, J. Cheng and S. Gedevanishvili, *Nature*, **399**, 668 (1999). (c) D. E. Clark and W. H. Sutton, *Annu. Rev. Mater. Sci.*, **26**, 299 (1996). (d) M. Gupta and W. Wong, *Scr. Mater.*, **52**, 479 (2005).

[9]Z. Radivojevic, K. Andersson, K. Hashizume, M. Heino, M. Mantysalo, P. Mansikkamaki, Y. Matsuba and N. Terada, *in Proc. 12th Intl. Workshop on Thermal investigations of ICs*, Nice, France, 2006.

[10]I. Reinhold, C. E. Hendriks, R. Eckardt, J. M. Kranenburg, J. Perelaer, R. R. Baumann, U. S. Schubert, *J. Mater. Chem.*, **19**, 3384 (2009).

[11]Z. Liu, Y. Su, K. Varahramyan, *Thin Solid Films*, **478**, 275 (2005).

[12](a) D. Hegemann, U. Vohrer, C. Oehr, R. Riedel, *Surf. Coat. Techn.*, **116-119**, 1033 (2005). (b) E. Vassallo, A. Cremona, L. Laguaria, E. Mesto, *Surf. Coat. Techn.*, **200**, 3035 (2006). (c) K. Teshima, Y. Inoue, H. Sugimura, O. Takai, *Chem. Vap. Deposition*, **8**, 251 (2002). (d) J. Kim, K. Lee, K. Kim, H. Sugimura, O. Takai, Y. Wu, Y. Inoue, *Surf. Coat. Techn.*, **162**, 135 (2003).

[13](a) F. H. Lu, H. D. Tsai, Y. C. Chieh, *Thin Solid Films*, **516**, 1871 (2008). (b) R. Schennach, T. Grady, D. G. Naugle, J. R. Parga, H. McWhinney, D. L. Cocke, *J. Vac. Sci. Technol. A*, **19**, 1965 (2001).

[14]S. Mathur, R. Ganesan, I. Grobelsek, H. Shen, T. Ruegamer, S. Barth, *Adv. Eng. Mat.*, **9**, 658 (2007).

[15](a) X. Hu, Y. Masuda, T. Ohji, K. Kato, *Langmuir*, **24**, 7614 (2008). (b) R. Kitsomboonloha, S. Baruha, M. T. Z. Myurit, V. Subramanian, J. Dutta, *J. Cryst. Growth*, **311**, 2352 (2009).
[16]B. Postels, M. Kreye, H.-H. Wehmann, A. Bakin, N. Boukos, A. Travlos, A. Waag, *Supperlattices Microstruct.*, **42**, 425 (2007).

NANOSTRUCTURED TIN DIOXIDE AND TUNGSTEN TRIOXIDE GAS SENSORS PREPARED BY GLANCING ANGLE DEPOSITION

Derya Deniz, Aravind Reghu, Robert J. Lad
Laboratory for Surface Science & Technology, University of Maine
Orono, ME 04469-5708, U.S.A.

ABSTRACT

Nanostructured tin dioxide and tungsten trioxide chemiresistive gas sensors were fabricated using a glancing angle deposition technique. The sensing films were grown using pulsed direct current magnetron sputtering at room temperature with continuous substrate rotation onto sensor platforms compromised of platinum electrodes, a thin film heater, and a resistance temperature device integrated onto a sapphire wafer. Structural and morphological properties of the films were characterized by X-ray diffraction and high resolution scanning electron microscopy. Both as-deposited and post-deposition air annealed tin dioxide and tungsten trioxide exhibit nanorod morphology with extremely high surface to volume ratios. These nanostructured sensors were exposed to 25 ppm ethylene gas and their response was compared to gas sensors fabricated using conventional normal incidence film deposition. The nanorod morphology induces a substantial increase in sensitivity. A thin gold catalyst layer deposited onto the sensors was found to markedly increase the surface reactivity towards ethylene and yield high sensitivity, especially for tungsten trioxide sensors.

INTRODUCTION

Tin dioxide (SnO_2) and tungsten trioxide (WO_3) are n-type semiconductors with wide band gap energies of 3.6 and 3.2 eV, respectively[1,2]. Their electrical conductivity can be varied significantly by controlling the number of oxygen vacancies in the crystalline lattice through interaction with the gaseous environment. Therefore, they have been widely used in gas sensing applications as conductance type chemiresistive sensors[3-10]. Thin films of metal oxides are conventionally grown by radio frequency (RF) magnetron sputtering to avoid target poisoning. Pulsed direct current (DC) sputtering can also be employed during sputtering of insulating materials, which allows the user to control the energy of reactive species more uniformly. This better control arises in DC sputtering because the ion energies in RF plasmas are distorted due to comparable time scales of the RF cycle and the ion trajectories across the sheath regions around both the substrate and the target[11].

Glancing angle deposition (GLAD) is a physical vapor deposition technique that can be used to engineer nanostructured films with morphological features such as nanorods, helixes, zigzags, etc. by adjusting the deposition angle, α and substrate rotation angle, φ[12,13]. The deposition angle, α, is usually kept at greater than 80° to take advantage of atomic shadowing effects. It is well known that high melting point materials have relatively low surface mobilities during film growth processes. Therefore, during a GLAD process, a high melting point material flux is chosen to be incident onto a substrate surface from an off-normal direction. Kinetic limitations such as shadowing and geometrical confinements coupled with low surface mobilities force the adatoms to form porous columnar microstructures[14,15]. Nanostructured thin films with their high surface to volume ratio are potentially very attractive for gas sensing applications since their unique morphology can improve sensor properties, namely: sensitivity, selectivity, response and recovery times[16,17]. The GLAD technique is very easy to incorporate into most sputter deposition chambers, and the films are easy to grow.

The sensing mechanism of chemiresistive sensors is based on the proposition that oxygen is chemisorbed around the grain boundaries on the film surface. Adsorbed oxygen species can be in the form of O_2, O_2^-, O^-, and O^{2-}, of which O^- has been shown to be the most reactive[18]. During reduction (oxidation) processes, oxygen ions are removed (added) from (onto) the film surface and subsurface resulting in a decrease (increase) in film resistance. Lattice oxygen O_L^{2-} may also react with the incoming reducing species. One can easily measure the film electrical resistance under a target gas exposure, thereby utilizing the film as an 'electronic nose'. The selective detection of a variety of alkanes in the environment at very low concentrations, such as methane (CH_4), ethylene (C_2H_4), ethane (C_2H_6), and propane (C_3H_8), are a challenge for chemiresistive sensors and is very relevant for a number of technical reasons.

In this study, we report the results of an investigation on nanostructured SnO_2 and WO_3 thin films grown using GLAD, and compare their gas sensing characteristics with relatively smooth SnO_2 and WO_3 films grown by conventional normal incidence magnetron sputtering. The influence of film structure and morphology, as well as post-deposition annealing treatments and addition of Au catalysts, on sensor characteristics are demonstrated for controlled exposures to C_2H_4 gas.

EXPERIMENTAL DETAILS

Nanostructured films were first deposited onto clean 2.5 cm x 2.5 cm fused quartz substrates under a variety of conditions in order to carry out structural and morphological analysis studies. Then, chemiresistive sensors were fabricated on r-cut sapphire platforms that consisted of either SnO_2 or WO_3 films on top of interdigitated platinum (Pt) electrodes on the front side, and with a serpentine heater and resistance temperature device (RTD) on the back side (see Fig.1b)[19]. The nanostructured films were grown in a high vacuum chamber with a base pressure of $< 1\times10^{-5}$ Pa by means of a home-made GLAD manipulator, which was coupled with a DC motor to sustain azimuthal rotation at a rate of 5 rpm during deposition. Fig.1a shows the specific GLAD geometry used for film growth.

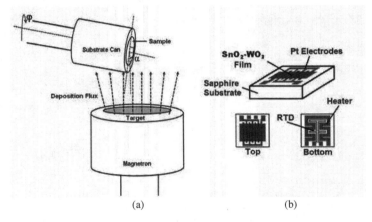

Figure 1. (a) Schematic diagram of the GLAD geometry used for nanostructured film growth.
(b) Chemiresistive gas sensor device on a 6 mm x 6 mm sapphire substrate.

The nanostructured films were grown at room temperature and at a deposition pressure of 0.15 Pa with Ar and O_2 flow rates of 4 and 6 sccm, respectively, using pulsed DC magnetron sputtering. The sputtering targets were 99.95% pure W or Sn 7.5 cm diameter disks placed 11 cm from the centerline of the substrate, and were presputtered in Ar for 10 min before each run. Nominal film thicknesses measured during deposition by a quartz crystal oscillator were determined to be 60 and 100 nm for SnO_2 and WO_3, respectively. Because of the nanorod morphology, the film thicknesses were more accurately determined by cross sectional scanning electron microscopy (SEM) measurements.

The SnO$_2$ films were deposited at a DC power of 30 W and a 436 V discharge voltage, whereas the WO$_3$ films were grown at a higher DC power of 100 W and a 410 V discharge voltage. The average deposition rates of SnO$_2$ and WO$_3$ were 0.9 and 3.0 nm/min, respectively.

To contrast with the nanostructured films, smooth polycrystalline WO$_3$[19] and SnO$_2$[20] films were grown on the sapphire sensor platforms using conventional normal incidence RF magnetron sputtering. In this case, an RF power of 54 W and a discharge voltage of 123 V were used at room temperature for SnO$_2$ fabrication whereas WO$_3$ fabrication required an RF power of 100 W and a discharge voltage of 100 V at 500°C substrate temperature.

Films were characterized right after deposition and also after annealing treatments up to 500°C in air. Film analysis consisted of X-ray diffraction (XRD) θ-2θ scans and grazing incidence angle scans using achromatic Cu K$_\alpha$ line focus radiation (λ_{avg} = 1.5406 Å) on a Panalytical X'pert MRD Pro Diffractometer. Sample morphologies were examined using a Zeiss Nvision 40 SEM on random areas of the films.

The chemiresistive sensors were tested in a home-made mass flow controller based gas delivery system. The system is computer supported and equipped with a KEPCO current source to heat the sensors, a KEITHLY 2000 multimeter, a 2400 source meter, and data logger software. Fig.2 shows a schematic of the gas delivery system. For the experiments reported here, VOC free dry air was flowed at a rate of 100 sccm over the sensors at a sensor operating temperature of 350°C. During exposure to the target gas, 25 ppm ethylene in VOC dry air carrier gas was switched to flow into the sensor test cell. In some cases, a thin Au catalyst layer (2 nm nominal thickness) was deposited by e-beam evaporation onto the sensors to increase the surface reactivity.

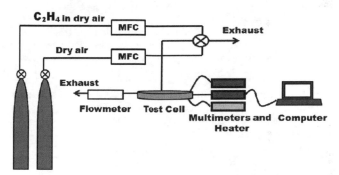

Figure 2. Schematic of the gas delivery system used to dose
25 ppm ethylene gas to the chemiresistive sensors.

RESULTS : FILM STRUCTURE AND MORPHOLOGY

As-grown SnO$_2$ and WO$_3$ films deposited by the GLAD technique exhibited amorphous structure as verified by XRD θ-2θ scans (not shown). This was also confirmed by grazing incidence angle XRD scans, which provide higher sensitivity to the film structure because of the grazing geometry.

A post-deposition annealing treatment in air at 400°C for 4 h caused the SnO$_2$ film to crystallize into a polycrystalline tetragonal (rutile) phase, which was verified by the data provided in the Joint Committee of Powder Diffraction Standards (JCPDS) card No. 01-071-5328. Fig.3 shows the grazing incidence XRD pattern from this annealed SnO$_2$ film. The grazing incidence pattern is sensitive to planes inclined with respect to the film surface, in contrast to θ-2θ scans which probe only planes parallel to the surface, and therefore many (hkl) orientations are visible from the polycrystalline film. The (200) and (120) peaks are much narrower than the other peaks suggesting that the grain size for these orientations is much larger than for the other grains. Since the (200) and (120) diffraction peaks occur at nearly the same 2θ positions as those from a tetragonal SnO phase, the presence of this coexisting phase cannot be ruled out.

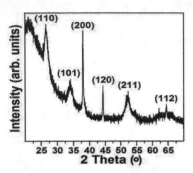

Figure 3. Grazing incidence angle XRD pattern from a nanostructured SnO₂ film that was post-deposition annealed in air at 400 °C for 4 h. The diffraction peaks are indexed to a polycrystalline tetragonal (rutile) structure.

The WO₃ film crystallized into a polycrystalline monoclinic and/or triclinic phase following an annealing treatment in air at 500 °C for 5 h as shown by the grazing incidence XRD pattern in Fig.4. Because the XRD peaks are broadened due to small grain size, the XRD peak positions are indistinguishably matched to both triclinic and monoclinic WO₃ phases given by the JCPDS cards No. 00-032-1395 and 01-083-0950. The exact phase identification in thin film WO₃ samples is always problematic because the lattice parameters of the several different WO₃ phases that have the distorted rhenium oxide structure are very close to each other and differ only by small amounts due to distortions of the WO₆ octahedral building blocks within the crystal lattice, especially when the structure contains oxygen vacancies[19].

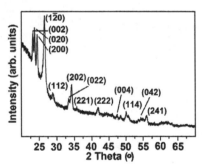

Figure 4. Grazing incidence angle XRD pattern from a nanostructured WO₃ film that was post-deposition annealed in air at 500 °C for 5 h. The diffraction peaks are indexed to a polycrystalline monoclinic/triclinic structure.

Fig.5a shows a plan view SEM micrograph from an as-deposited SnO_2 film deposited with the GLAD technique. Despite the fact that the film is amorphous, the film is composed of nanorods with average diameter of ~ 15 nm protruding from the surface. These nanorods are separated from each other, leading to a highly porous morphology with a very large surface to volume ratio. The annealing treatment in air at 400°C did not significantly alter the nanomorphology of the SnO_2 film as seen in Fig.5b. These results suggest that these nanostructured SnO_2 films will remain very stable during gas sensor operation at elevated temperatures.

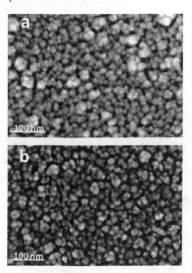

Figure 5. Plan view SEM images of (a) an as-deposited SnO_2 GLAD film and
(b) a SnO_2 GLAD film after post-deposition annealing treatment
in air at 400 °C for 4 h.

Fig.6a shows a plan view SEM image from an as-deposited WO_3 film deposited with the GLAD technique. As with the SnO_2 film, nanorods are present in the film, but in this case several nanorods agglomerate together to form so-called nanostructured cauliflower morphology, which is a natural consequence of a GLAD process during its later stages. In contrast to SnO_2, the post-deposition annealing treatment at 500°C causes a very significant morphological change into a connected nanoporous network (Fig. 6b). The large interconnected grains in this network are consistent with the more narrow XRD lineshapes in Fig.4 for a larger grained WO_3 film. The fact that the surface mobilities of the SnO_2 and WO_3 deposition fluxes are different, and that the WO_3 film used for the SEM analysis was ~40 nm thicker than the SnO_2 film, could explain the cauliflower structure in the WO_3 case. For this reason, the SnO_2 and WO_3 films deposited on the sensor platforms, and discussed in the next section, were both grown to an identical film thickness of ~ 50 nm.

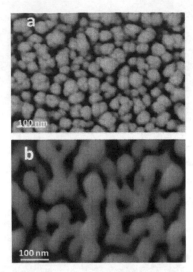

Figure 6. Plan view SEM images of (a) an as-deposited WO₃ GLAD film and
(b) a WO₃ GLAD film after post-deposition annealing treatment
in air at 500 °C for 5h.

RESULTS : FILM GAS SENSING CHARACTERISTICS

Nanostructured WO₃ and SnO₂ films prepared by the GLAD technique as well as flat polycrystalline WO₃ and SnO₂ films prepared by conventional normal incidence RF magnetron sputtering were deposited onto sapphire chemiresistor platforms. The films were all nominally 50 nm thick. Ethylene gas (C_2H_4) was chosen as the target gas to compare the sensor response of each of the films. During each gas sensing experiment, the sensor was stabilized for a least one hour at the 350°C operating temperature in VOC free dry air flowing at 100 sccm. The established base line resistance, R, ranged from low to high mega-ohm values depending on the film type. During target gas exposure, 25 ppm C_2H_4 was introduced into the gas stream and the change in resistance, ΔR, was measured from the sensor. After a 45 min exposure, pure dry air was introduced into the sensor test cell and the sensor resistance recovered to the initial base line value. The sensitivity of the sensor, S, towards ethylene gas was determined from the ratio ΔR / R.

Table I shows the cumulative results of sensor tests to the various nanostructured and flat WO₃ and SnO₂ films. Ethylene acts as a reducing gas and therefore ΔR is negative for each sensor. For the WO₃ case, the nanoporous structured film yielded a two-fold sensitivity increase over the flat polycrystalline film. The sensitivity of both films could be further enhanced by adding a Au catalyst layer to the surface. In particular, the Au-doped WO₃ GLAD film exhibited a very rapid response and high sensitivity (0.81) as shown in Fig.7. The higher sensitivity compared to the flat WO₃ film indicates the effectiveness of the large film surface to volume ratio.

In the case of SnO₂, the undoped films are already very sensitive to ethylene gas. No significant difference is observed between the nanostructured and flat films, presumably since the surface reactivity is already very high, and an increase in film surface area has a minimal effect. Additional sensor response studies with other gases are underway to further investigate these effects and will be reported in a future publication.

Table I. Comparison of sensor response of different film types to 25 ppm C_2H_4 exposure

Sample Type	ΔR (MΩ)	R (MΩ)	Sensitivity = $\Delta R/R$
WO$_3$ Flat	-1.6	14.5	0.11
WO$_3$ GLAD	-0.46	2.06	0.22
Au - WO$_3$ Flat	-0.017	0.069	0.25
Au - WO$_3$ GLAD	-59	72.5	0.81
SnO$_2$ Flat	-6.9	7.5	0.92
SnO$_2$ GLAD	-0.14	0.16	0.88

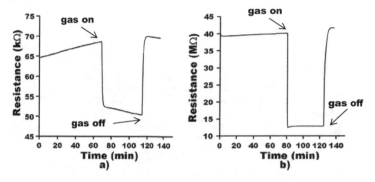

Figure 7. Sensor response to 25 ppm C_2H_4 in dry air at 350°C operating temperature. (a) flat polycrystalline WO$_3$ film doped with Au and (b) a WO$_3$ GLAD film doped with Au. Note the different resistance scales on each plot.

CONCLUSIONS

Glancing angle deposition (GLAD) of SnO$_2$ and WO$_3$ films is an effective way to produce chemiresistive gas sensing films with large surface to volume ratios. The low surface mobilities of the sputter deposition flux combined with shadowing effects during film growth lead to nanorod film morphologies. Growth of SnO$_2$ and WO$_3$ films by GLAD at room temperature yields amorphous films, but post-deposition annealing treatments up to 500°C cause crystallization into tetragonal SnO$_2$ or monoclinic/triclinic WO$_3$ phases. The crystallized nanostructured WO$_3$ films exhibited a higher sensitivity towards ethylene gas compared to the conventionally sputter deposited flat polycrystalline films. Adding a gold catalyst further enhances the sensitivity for both film types. In the case of SnO$_2$, the high surface reactivity toward ethylene induced a high sensitivity regardless of the film type, indicating that morphology effects were less important in this case. Additional sensor testing studies with other target gases are currently being carried out to further investigate the influence of film morphology on sensor response.

ACKNOWLEDGEMENTS

This work was financially supported by the W.M. Keck Foundation. The authors are grateful to G. Bernhardt, M. Call, D. Frankel, and E. Martin of the University of Maine for their help with the gas delivery system.

REFERENCES

[1]C. A. Papadopoulos and J. N. Avaritsiotis, A Model for The Gas Sensing Properties of Tin Oxide Thin Films with Surface Catalysts, *Sens. and Actuators B*, **28**, 201-210 (1995).

[2]A. Raugier, F. Portemer, A. Quede, M. ElMarssi, *Appl. Surf. Sci.*,**153**, 1-9 (1999).

[3]M. Ippommatsu, H. Ohnishi, H. Sasaki, and T. Matsumoto, Study on Sensing Mechanism of Tin Oxide Flammable Gas Sensors Using the Hall Effect, *J. Appl. Phys.*, **69**, 8368-8374 (1991).

[4]G. Tournier, C. Pijolat, R. Lalauze, and B. Patissier, Selective Detection of CO and CH_4 with gas sensors using SnO_2 doped with Palladium, *Sens. and Actuators B*, **26-27**, 24-28 (1995).

[5]S. H. Kim, K. T. Lee, S. Lee, J. H. Moon, and B. T. Lee, Effects of Pt/Pd Co-Doping on the Sensitivity of SnO_2 Thin Film Sensors, *Jpn. J. Appl. Phys.*, **41**, L1002-L1005 (2002).

[6]T. Sahm, W. Rong, N. Barsan, L. Madler, and U. Weimar, Sensing of CH_4, CO and Ethanol with in situ Nanoparticle Aerosol-Fabricated Multilayer Sensors, *Sens. and Actuators B*, **127**, 63-68 (2007).

[7]M. Stankova, X. Vilanova, E. Llobet, J. Calderer, M. Vinaixa, I. Gracia, C. Cane, X. Correig, On-line Monitoring of CO_2 Using Doped WO_3 Thin Film Sensors, *Thin Solid Films*, **500**, 302-308 (20076).

[8]Y. G. Kim, Thermal Treatment Effects on the Material and Gas Sensing Properties of Room-Temperature Tungsten Oxide Nanorod Sensors, *Sens. Actuators B* **137** 297-304 (2009).

[9]A. Ponzoni, E. Comini, G. Sberveglieri, J. Zhou, S. Z. Deng, N. S. Xu, Y. Ding, and Z. L. Wang, Ultrasensitive and Highly Selective Gas Sensors Using Three-Dimensional Tungsten Oxide Nanowire Networks, *Appl. Phys. Lett.* **88**, 203101-3 (2006).

[10]E. H. Williamson and N. Yao, Tungsten Oxide Nanorods: Synthesis, Characterization, and Application, In: B. Zhou, S. Han, R. Raja, and G. J. Somorjai, *Nanostructure Science and Technology, Nanotechnology in Catalysis* 3, New York: Springer, 115-137 (2007).

[11]E. Barnat and T. M. Lu, Pulsed Bias Magnetron Sputtering of Thin Films on Insulators, *J. Vac. Sci. Technol. A* 17 3322-3326 (1999).

[12]K. Robbie and M. J. Brett, Sculptured Thin Films and Glancing Angle Deposition: Growth Mechanics and Applications, *J. Vac. Sci. Technol. A* **15**, 1460-1465 (1997).

[13]K. Robbie, J. C. Sit, and M. J. Brett, Advanced Techniques for Glancing Angle Deposition, *J. Vac. Sci. Technol. B* **16**, 1115-1122 (1998).

[14]M. M. Hawkeye and M. J. Brett, Glancing Angle Deposition: Fabrication, Properties and Applications to Micro- and Nanostructured Thin Films, *J. Vac. Sci. Technol. A* **25**, 1317-1335 (2007).

[15]L. Abelmann and C. Lodder, Oblique Evaporation and Surface Diffusion, *Thin Solid Films* **305**, 1-21 (1997).

[16]A. Kolmakov and M. Moskovits, Chemical Sensing and Catalysis by One-Dimensional Metal-Oxide Nanostructures, *Annu. Rev. Mater. Res.*, **34**, 151-180 (2004).

[17]E. Comini, Metal Oxide Nano-Crystals for Gas Sensing, *Anal. Chim. Acta*, **568**, 28-40 (2006).

[18]N. Yamazoe, J. Fuchigami, M. Kishikawa, and T. Seiyama, Interactions of Tin Oxide Surface with O_2, H_2O and H_2, *Surf. Sci.*, **86**, 335-344 (1979).

[19]S.C. Moulzolf, L.J. LeGore, and R.J. Lad, Heteroepitaxial Growth of Tungsten Oxide Films on Sapphire for Chemical Gas Sensors, *Thin Solid Films* **400**, 56-63 (2001).

[20]R.E. Cavicchi, S. Semancik, M.D. Antonik, and R.J. Lad, Layer-by-layer Growth of Epitaxial SnO_2 on Sapphire by Reactive Sputter Deposition, *Appl. Phys. Lett.* **61**, 1921-1923 (1992).

HYDROTHERMAL SYNTHESIS OF TIO$_2$ NANOTUBES: MICROWAVE HEATING VERSUS CONVENTIONAL HEATING

Lucky M. Sikhwivhilu[1*,] Siyasanga Mpelane[1], Nosipho Moloto[1,2], Suprakas Sinha Ray[1]

[1]DST/CSIR Nanotechnology Innovation Centre, National Centre for Nano-Structured Materials, Council for Scientific and Industrial Research, P. O. Box 395, Pretoria, 0001, Republic of South Africa.
[2]Molecular Sciences Institute, School of Chemistry, University of the Witwatersrand, Private Bag 3, Wits, 2050, Republic of South Africa.

ABSTRACT
    The influence of the method of synthesis in the properties of the tubular structures derived from TiO$_2$ was investigated using XRD, SEM and BET analysis. The use of microwave irradiation resulted in the formation of TiO$_2$ tubes comprising anatase and rutile phases. Conventional heating resulted in the formation of tubes with a titanate structure. The two methods yielded tubular structures with similar size dimensions, surface areas and morphologies. The two methods gave 100 % yields of tubes with different degrees of crystallinity.

INTRODUCTION

    Synthesis and engineering of nanostructured semiconductors based on metal oxides have received considerable attention due to their unique physical and chemical properties, and their potential applications in industry and technology [1-2]. Whilst various methods have been used to synthesize TiO$_2$ nanoparticles, the hydrothermal synthesis in the presence of a base solution, has proved to be an effective approach to prepare 1D nanostructures of TiO$_2$. This is because the method utilizes minimum reagents and produces relatively purer materials. However, the main attention is directed towards controlling the structure and morphology of the particles by varying the synthesis conditions such as temperature, pressure and time of processing during hydrothermal processing [3-4].
    The use of microwave irradiation in the synthesis of nanomaterials is becoming an important tool to fabricate materials with specific properties [5]. The fact that energy is delivered to the reactants through molecular interactions with electromagnetic field holds promise of improved synthesis [6-7]. It can provide accelerated reaction rates, high energy density and short reaction times leading to nanoparticles with improved crystallinity. It is envisaged that this would allow for the control of particle size, degree of crystallinity and morphology [8-11].

    In our study, TiO$_2$ nanostructures are synthesized using conventional heating and microwave-assisted hydrothermal procedures. The effects of heating on the size, shape and crystallinity of materials are studied. Microwave heating is particularly interesting because of the utilization of higher energy density and shorter reaction times leading to nanoparticles that are weakly agglomerated, with high crystallinity and narrow particle size distribution.

EXPERIMENTAL

Synthesis

The synthesis of TiO$_2$ derived nanotubes (TNT) was carried out using a procedure described elsewhere [12-13]. In a typical procedure about 23 g of TiO$_2$ powder, P25 Degussa, was mixed with 200 ml of 18 M of aqueous solution of KOH. The mixture was either heated at 150 °C in an autoclave with constant stirring for 24 hours or irradiated in a microwave reactor for 15 minutes with the power of 600 W and maximum temperature set at 150 °C. The products were then washed with deionised water and then dried in an oven at 120 °C for 14 hours. The sample generated by using conventional heating in the autoclave was designated TNT-A, whereas the sample obtained by microwave irradiation was designated TNT-B.

Characterization

The surface area measurements were taken using a BET Tristar III instrument. XRD patterns on powdered samples were measured on a Phillips X'Pert materials research diffractometer using secondary graphite monochromated Cu Kα radiation (λ = 1.54060 Å) at 40 kV/50 mA. Measurements were taken using a glancing angle of incidence detector at an angle of 2°, for 2θ values over 10° – 90° in steps of 0.05° with a scan speed of 0.01° 2θ.s^{-1}. The SEM measurements were carried out in a Leo, Zeiss FE-SEM microscopy operated at 2 kV electron potential difference and equipped with a semiconductor detector that allows for detection of energy dispersive X-rays (EDX).

RESULTS AND DISCUSSION

It is noteworthy to realise that the two samples TNT-A and TNT-B were obtained under optimum conditions for each method. TNT-A showed a specific surface area of 246 m^2/g whereas TNT-B showed the specific surface area of 240 m^2/g. The specific surface areas obtained for both materials are comparable showing that the surface area is not largely influenced by the method of synthesis and can attest to the similar morphologies of the products.

Table 1: Physical properties of TNT-A and TNT-B

Sample	Heating Method	Oven Temperature (°C)	S$_{BET}$ (m^2/g)
TNT-A	Conventional	150	245.9
TNT-B	Microwave	150	239.9

The crystal phases of both TNT-A and TNT-B were characterised by XRD as shown in **Fig. 1**. The XRD pattern of TNT-A shows the presence of fewer and broader peaks than TNT-B. The peaks could not be indexed to either the anatase or rutile phase. However, from our previous results and exhaustive literature search the peaks were found to correspond to a titanate structure, KTiO$_2$(OH) [13-17]. The XRD pattern of TNT-B shows the presence of sharper peaks which could be indexed to both anatase and rutile phases. Though the material is mainly comprised of anatase and rutile a reasonable amount of titanate structure, Ti$_3$O$_7$, was also found to be present. The phase identification for anatase with XRD was based on (1001) (004) (200) (105) (211) (204) and (116) peaks at 25.34, 37.81, 48.10, 53,92, 55.14,62.75 and 68.81 2-theta degrees respectively [18] whereas the rutile phase identification was based on (110) (101) (111) (211) (220) and (301) peaks at 27.45, 36.09, 41,23, 54.32, 56,64, and 69.01 2-theta degrees respectively [19]. The titanate peak is more prominent at 2-theta value of about 10 degrees. Through semi-quantitative XRD analysis, the volumetric fractions of the anatase, rutile and titanate phases in TNT-B were estimated to be about 35, 41 and 24%, respectively. This clearly shows that the tubes in this sample are of TiO$_2$ (76%) and Ti$_3$O$_7$ (24%) forms and that anatase, rutile and titanate phases co-exist in TNT-B. The crystallinity of the sample synthesized using microwave irradiation (TNT-B) was found to be better than that of the sample prepared by conventional heating (TNT-A) evident from the peak sharpness. This is attributed to the difference in structural composition and possibly the short reaction time for microwave processing.

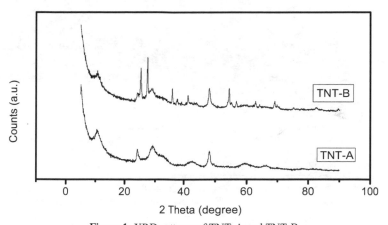

Figure 1: XRD patterns of TNT-A and TNT-B

The SEM results of the samples prepared using conventional heating (TNT-A) and microwave heating (TNT-B) are shown in **Fig. 2**. Both samples are composed of tubular structures with diameter range of 8-11 nm. Though the methods of synthesis and conditions for both TNT-A and TNT-B are different, the size dimensions are comparable. TNT-A formed relatively more and bigger bundled tubular structures than TNT-B. This implies that TNT-A is more agglomerated than TNT-B. This is ascribed to the difference in reaction times and surface charges. The presence of K$^+$ ions in the TNT-A sample could lead to increased surface charge and electrostatic attraction of individual tubes subsequently leading to agglomeration. The EDX results (not shown) of TNT-B did not reveal the presence of potassium indicating that the tubes obtained by microwave processing are TiO$_2$ and Ti$_3$O$_7$.

Even though the dimensions and morphologies of the tubes are similar, the structural compositions are different. This suggests a possibility of different growth mechanisms. The difference in the required times of processing is evident of the mechanisms path being either kinetically or thermodynamically controlled. It is therefore envisaged that the conventional heating method follows the thermodynamic pathway that result in the formation of KTiO$_2$(OH) and that TiO$_2$ and Ti$_3$O$_7$ nanotubes form under kinetic conditions. From this observation the relative stability of the products can be deduced. The kinetically formed products are usually less stable compared to the thermodynamically formed products and this would explain the co-existence of anatase, rutile and titanate phases in the microwave synthesized TiO_2 derived nanostructures. However TiO_2 (probably without Ti_3O_7 phase) nanotubes are more desirable than $KTiO_2(OH)$ nanotubes due to their potential applications. The microwave process with further manipulation; as a result of its rapid growth rate, can therefore result in the attainment of relatively purer (structures made up of only Ti and O atoms) tubular structures.

Figure 2: SEM images of (a) TNT-A and (b) TNT-B samples.

CONCLUSION

The results showed that the TiO_2 derived tubes obtained by conventional heating and microwave irradiation differ in their structural and phase composition. TNT-A (obtained by conventional heating) showed a titanate structure whilst TNT-B (obtained by microwave irradiation) gave a TiO_2 structure with anatase, rutile and titanate phases in different proportions.

The results also showed that microwave irradiation leads to rapid heat generation in the sample that the morphological transformation occurs without the reduced structural transformation. Subsequently, this leads to the formation of material with the co-existence of anatase, rutile and titanate phases. Although the two methods employed for the synthesis of TiO_2 derived nanotubes yielded tubes with similar morphology, size dimensions, and surface area the molecular structure, crystal structure and crystallinity were influenced by experimental conditions and method of synthesis. Tubes synthesized by long processing procedure using conventional heating (TNT-A) tended to agglomerate easily due to surface charge accumulation.

REFERENCES

1. B.D. Yao, Y.F. Chan, X.Y. Zhang,W.F. Zhang, Z.Y. Yang, N.Wang, Appl. Phys. Lett. 82 (2003) 281.
2. S. Zhang, Q. Chen, L.M. Peng, Phys. Rev. B 71 (2005) 014104.
3. S.H. Kang, J.Y. Kim, Y. Kim, H.S. Kim, Y.E. Sung, J. Phys. Chem. C 111 (2007) 9614.
4. B. Tan, Y.Wu, J. Phys. Chem. B 110 (2006) 15932.
5. S.A. Freeman, J.H. Booske, R.F. Cooper, J. Appl. Phys. 83 (1998) 5561.
6. H.X. Liu, Y.W. Li, H.L. Zhang, Sci. China, Ser. A 40 (1997) 843.
7. Y.W. Li, H.X. Liu, H.L. Zhang, S.X. Ouyang, Sci. China Ser. A 40 (1997) 779.
8. D.E. Clark, D.C. Folz, Mater. Res. Soc. Symp. Proc. 347 (1994) 489.
9. H.X. Liu, Z.J. Liu, S.X. Ouyang, Acta Phys.-Chim. Sinica 14 (1998) 624.
10. H.X. Liu, Z.J. Liu, S.X. Ouyang, Acta Chim. Sinica 57 (1999) 472.
11. M. Fu, Z.D. Jiang, Z.F. Ma, W.F. Shangguan, J. Inorg. Mater. 20 (4) (2005) 808.
12. T. Kasuga, M. Hiramatsu, A. Hoson, T. Sekino, K. Niihara, Langmuir 14 (1998) 3160.

13. L.M. Sikhwivhilu, S. Sinha Ray, and N.J. Coville, Appl. Phys. A 94 (2009) 963.
14. L.M. Sikhwivhilu, N.J. Coville, D. Naresh, K.V.R. Chary, V. Vishwanathan, Appl. Catal. A: Gen. 52 (2007) 324.
15. J. Yang, Z. Jin, X. Wang, W. Li, J. Zhang, S. Zhang, X. Guo, Z. Zhang, Dalton Trans. (2003) 3898. doi: 10.1039/b305585
16. N. Masaki, S. Uchida, H. Yamana, T. Sato, Chem. Mater. **14**, 419 (2002)
17. L.M. Sikhwivhilu, N.J. Coville, B.M. Pulimaddi, J. Venkatreddy, V. Vishwanathan, Catalysis Communications 8 (2007) 1999.
18. E. Sanchez, T. López, R. Gómez, M.A. Bokhimi, O. Novaro, J. Solid State Chemistry 122 (1996) 309.
19. Natl. Burg. Stand. Monogr. "Crystallography data of rutile", USA, 25 (1969) 83.

FABRICATION AND PROPERTIES OF CORE-SHELL TYPE SiC/SiO$_2$ NANOWIRES THROUGH LOW-COST PRODUCTION TECHNIQUE

Wasana Khongwong, Katsumi Yoshida and Toyohiko Yano

Research Laboratory for Nuclear Reactors, Tokyo Institute of Technology,
2-12-1, O-okayama, Meguro-ku, Tokyo 152-8550, Japan

ABSTRACT

A worth challenge in study on synthesis of SiC nanowires is to seek a low-cost and an un-complex production method for synthesis a large amount of SiC nanowires. In this study, the simple and inexpensive production route to fabricate nanowires through the reaction of cheap Si powder and CH$_4$ gas in a tube furnace, using separate and continuous heating process, was reported. This method, thermal evaporation, can produce SiC crystalline core/SiO$_2$ low crystallinity shell composite-phase nanowires at a processing temperature of 1350°C. The results show that both of separate heating process (performed with Si powders of 5 μm in average particle size) and continuous heating process, (performed with Si powder of 74 μm in average particle size) could obtain SiC/SiO$_2$ core-shell nanowires. The typical synthesized nanowires owned with approximately 80 nm and 0.5-2 mm in diameter and length, respectively. Photoluminescence of synthesized nanowires showed two broad photoluminescence peaks located around 405 nm and 470 nm under 260 nm UV excitation at room temperature. Thermal conductivity of sintered pellets without pores prepared from Al$_2$O$_3$-SiC nanowire mixture occupied almost constant thermal conductivity at each amount of added-SiC nanowire. Thermal conductivity of sintered pellets was mostly dominated by density of pellets, both in nanowire-added and nanopowder-added system.

1. INTRODUCTION

During the fast development of nanotechnology in the past decade, SiC nanowires or nanocables have attracted considerable attention among many novel one-dimension nanomaterials since they are found to have various special properties and potential application.[1] SiC nanowires, as a wide bandgap semiconductor (2.3-3.2 eV), with high thermal conductivity (300-500 Wm^{-1}K^{-1}), high electron saturation velocity (2.0 x 10^5 ms^{-1}) and high resistant to chemical corrosion,[2] show potential for application under a range of harsh conditions including high-temperature, high power and high frequency.[3] Moreover, high strength, low density, high stiffness and high temperature stability combined with a high aspect ratio make 3C-SiC nanowires very effective reinforcement for various composites. The strength of SiC nanowire has been found to approach the theoretical strength and is substantially larger than that found in bulk SiC.[4] Wong et al.[5] estimated the yield strength of SiC nanowires (over 50 GPa) using atomic force microscopy. The morphology of nanomaterials is known to have influence on their properties.[6] In comparison with the whiskers, SiC nanowires have a larger aspect ratio and better elastic modulus and strength.[5] Therefore, SiC nanowires should be more suitable to be used as the reinforcing materials for ceramics than SiC whiskers.

SiO$_2$ is an insulator and SiO$_2$ nanotubes have been demonstrated to be highly valuable in bioanalysis, bioseparation, and optics.[7-9] SiC/SiO$_2$ nanowires, with crystalline SiC core and amorphous SiO$_2$ shell, are ideal semiconductor-insulator heterostructures in radial direction, and are expected to have excellent properties of both SiC nanowires and SiO$_2$ nanotubes. In addition, SiC nanowires can emit blue-green light,[10-12] so they would have great potential as light-emitting devices as well. Therefore, a lot of efforts have been made to synthesize SiC nanowires,[13-16] since research on both synthesis and properties of SiC nanowires is meaningful.

In ours previous works,[17, 18] a large amount of SiC/SiO$_2$ core-shell nanowires can be produced via reaction of evaporated Si, SiO gas and CH$_4$ gas, and effect of process parameters on amount of products was clarified. Here, not only synthesis of SiC/SiO$_2$ nanowires using oxidized Si powder or ground Si ingot as raw powders through separate and continuous processes, but also optical properties of as-grown products and thermal properties of nanowire-added Al$_2$O$_3$ based composites were investigated.

2. EXPERIMENTAL PROCEDURE

2.1 FABRICATION OF SiC/SiO$_2$ CORE-SHELL NANOWIRES

The fabrication of SiC/SiO$_2$ core-shell nanowires was similar to that of our previous work.[19] Briefly, SM (silicon powder; average particle size ≈ 5 μm, 99% nominal purity, Kojundo Chemical Laboratory Co., Ltd., Japan) was oxidized for 1 h in air at 800°C before put in a mullite boat which was then covered with an alumina fiber sheet (Nextel^{TM} Woven Fabric 610 Style, Sumitomo 3M Ltd., Japan). The whole set was carefully pushed into the middle of a tube furnace.

Before heating, the tube furnace was evacuated for 1.50 h to a pressure below 1.33 Pa using a mechanical rotary pump, then the ultra high purity Ar gas (purity: 99.9995%) was released into the furnace at a flow rate of 0.6 dm^3/min to reduce the oxygen to a negligible level. The furnace was initially raised to 1200°C at a heating rate of 10°C/min and then continued to heat to a peak temperature (1350°C) with heating rate of 5°C/min. At 1350°C, H$_2$ gas (purity: 99.999%) at a flow rate of 20 sccm (1 sccm = 1.667 x 10^{-8} m^3/s) was fed for 2 min before flowing of CH$_4$ gas at a flow rate of 10 sccm. CH$_4$ gas was fed for initial 30 min of keeping at 1350°C. The reaction was kept at the target temperature for 1 h. Finally, the furnace was cooled to room temperature under an Ar atmosphere. Synthesis process as mentioned above was named as separate heating process or H1, i.e., oxidation of raw powder and nanowires synthesis were separately conducted. The obtained products synthesized via this process was coded as SM/O8-H1.

To reduce cost and shorten time for the production, SG (silicon ingot; dark gray, 99% nominal purity, Hirano Seizaemon Co., Ltd., Japan) and continuous heating process (H2) instead of SM and H1 were used to prepare nanowires. H2 is continuous heating process, i.e., multi-step heating for oxidation at 800°C and reaction at 1350°C continuously in the same furnace. The Si ingot was ground and then sieved with a No. 200 mesh sieve (74 μm) before used as a precursor (names as SGG). The obtained products synthesized using SGG through H2 was coded as SGGO8-H2.

The as-grown nanowires and sintered pellets mentioned in section 2.3 were characterized by XRD (CuKα, PW 1700, Philips, the Netherlands), FE-SEM (Model S-4800, Hitachi, Japan), TEM (Model H-9000, Hitachi, Japan), and FT-IR (Model FT/IR-460 Plus, JASCO, Japan).

2.2 PHOTOLUMINESCENCE MEASUREMENT

The synthesized SiC/SiO$_2$ core-shell nanowires from SM/O8-H1 specimen was used as representative one to measure the optical property. Photoluminescence (PL) of SiC nanowires was measured by a fluorescence spectrophotometer (Hitachi, F-4500, Japan) using 260 nm excitation light from a xenon lamp at room temperature. Sample was compacted into the sample holder before put down in the analyzing chamber.

2.3 THERMAL CONDUCTIVITY MEASUREMENT

Submicron α-Al$_2$O$_3$ powder of 99.99% purity (Taimicron TM-D, average particle size 0.16 μm, Taimei Chemicals Co., Ltd, Japan) was mixed with a small amount of SiC/SiO$_2$ core-shell nanowires from SM/O8-H1 specimen. Three different amounts of SiC/SiO$_2$ nanowire, 0.035, 0.1 and 0.2 wt%, were mixed with the Al$_2$O$_3$ powder and named as ASw(0.035), ASw(0.1) and ASw(0.2), respectively. For comparison, the same three amounts of SiC nanopowder (average particle size ~ 30 nm, Sumitomo Osaka Cement Co., Ltd., Japan) were blended with the Al$_2$O$_3$ raw powder as well. The mixtures of Al$_2$O$_3$ and SiC nanopowder were named as ASp(0.035), ASp(0.1) and ASp(0.2), respectively. The mixture was ball-milled in a polyethylene bottle using silicon carbide balls ($\phi = 5$ mm) and ethyl alcohol as medium. After mixing for 24 h, the slurry was dried in an oven at a temperature of 120°C. Dried powder was uniaxially pressed into pellets under 5 MPa and was then isostatically cold-pressed (CIP) under a pressure of 200 MPa. Specimens (13 mm in diameter and 2.5 mm in thickness) were placed in a mullite boat and sintered in a mullite tube furnace at the temperature of 1350°C for 1 h in Ar atmosphere.

After sintering, mass loss during sintering process of Al$_2$O$_3$ discs was measured. Bulk density was obtained by the Archimedes method using water. Before measuring thermal conductivity, the sintered specimens were polished to reduce the dimension from about 2 mm to 1 mm in thickness and from about 11 mm to 10 mm in diameter and both sides of surface were then coated with dry graphite film lubricant (DGF, Nihonsempakukougyu, Japan). The thermal diffusivity (α) of the carbon-coated specimens were measured at room temperature by laser flash method (LF/TCM, FA 8510B, Rigaku, Japan). Thermal conductivity with pores (κ_p) was calculated according to the equation, $\kappa_p = \rho \cdot C_p \cdot \alpha$, where ρ is bulk density, C_p is specific heat and α is thermal diffusivity. Thermal conductivity without pores (κ) was estimated following the equation, $\kappa = \kappa_p/[(1-v_d)/(1+v_d)]$, where v_d is porosity of sample. Specific heat of pure alumina at room temperature (0.88 J/g.K)[20] was used to calculate thermal conductivity.

3. RESULTS AND DISCUSSION

3.1 CHARACTERIZATION OF NANOWIRES

After reaction, larger quantity of white-blue wool-like products was obtained on SM/O8-H1 surface (see Fig. 1 (a)) than that on SGGO8-H2 surface (see Fig. 1 (d)). SGGO8-H2 was a trial to

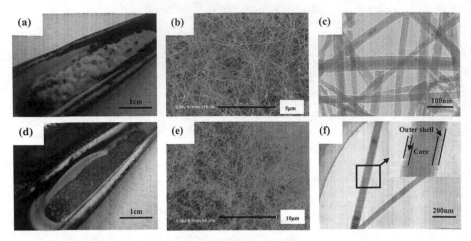

Fig. 1 (a) Photograph of as-grown product in a mullite boat from SM/O8-H1, (b) FE-SEM image of wool-like product from SM/O8-H1, (c) TEM image of typical nanowires from SM/O8-H1, (d) photograph of as-grown product in a mullite boat from SGGO8-H2, (e) FE-SEM image of wool-like product from SGGO8-H2, and (f) TEM image of typical nanowires from SGGO8-H2 (The upper right inset is enlarged image of a small square area.

prepare with the low-priced Si powder. Although the wool-like products from SGGO8-H2 was less than the wool-like products from SM/O8-H1, this way is one alternative process to synthesize wool-like products with cheaper raw material and shorter time to production.

XRD patterns of as-grown products (not shown here) both prepared from SM and SGG confirmed that crystalline phase of all the deposition products was β-SiC. There exist four main strong peaks which can be attributed to the (111), (200), (220) and (311) planes of the cubic type SiC phase. These 2θ or d values are almost identical with the known values for β-SiC (JCPDS Card No. 29-1129).

The typical FE-SEM images of nanowires prepared from SM/O8-H1 and SGGO8-H2 specimens were shown in Fig. 1 (b) and (e), respectively. Both the wool-like products from SM/O8-H1 and SGGO8-H2 specimens composed of a large amount of straight, curved, tangled, randomly distributed nanowires. Length of the nanowires synthesized from oxidized SM at 800°C was too long to be measured under FE-SEM, and it is estimated to be 1 to 2 mm from the height of the product grown on SM raw powder surface. Whereas the length of the nanowires from SGGO8-H2 specimen was about 0.5 mm.

TEM observation confirmed that a structure of as-grown nanowires both from SM/O8-H1 (see Fig. 1 (c)) and SGGO8-H2 (see Fig. 1 (f)) specimens was core-shell structure and the surface was very smooth. The diameter of core of nanowires was ranging from 20 to 80 nm, and it was wrapped with a uniform layer shell with a thickness of 10-20 nm. To obtain more details about the structure and

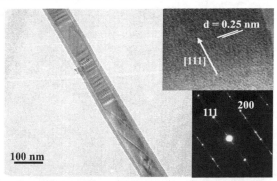

Fig. 2 TEM image of typical nanowires from SGGO8-H2 specimen. The upper right inset is a high magnification image of nanowire (The [111]-growth direction is indicated by an arrow). The lower right inset is the corresponding SAD pattern of SiC core.

crystallinity of synthesized nanowires, selected-area electron diffraction (SAD) methods were conducted and the results are shown in Fig. 2. The SAD pattern showed that the crystalline SiC core had stacking faults and twins. It is well known that the SiC nanowires typically occupy high density of stacking faults and twin defects.[21, 22] High-magnification image of SiC nanowire indicated that fringe of 0.25 nm-repeat corresponding to the d-spacing of the (111) plane. The growth direction of the nanowire was [111] of β-SiC, as same as indicated previously.[17, 18] Generally, it is accepted that β-SiC nanowires can grow easily in the [111] direction because the {111} surface have the lowest surface energy among the SiC surfaces and to decrease the formation energy, and hence stacking faults can be inserted easily in the (111) plane.[23]

Fig. 3 is the FT-IR spectrum of the composite (core/shell) nanowires obtained from SM/O8-H1 specimen. Two absorption bands from Si-O stretching vibration at around 1102 and 466 cm^{-1}, transversal optic (TO) mode of Si-C vibration at around 804 cm^{-1} and a shoulder which was marked by a small circle at around 860-950 cm^{-1} corresponding to the longitudinal optic (LO) mode of Si-C vibration were observed. The result is just in agreement with the previous reports.[24] Together with the XRD and TEM analyses, it is believed that the outer shell is consisted from low crystallinity or amorphous SiO$_2$.

In the growth process of the composite nanowires, the reaction is seemingly involved with SiO vapor phase. As the temperature increases, SiO vapor is generated by the reaction of SiO$_2$ thin layer on the surface of Si and evaporated Si. Dense SiO smoke is deposited first near Si powder surface. Subsequently, when the furnace is heated up till close to the melting point of Si (1414 °C), evaporation of Si accelerated. The main reaction of SiC nucleation might be SiO (v) + Si (v) + 3CH$_4$ = 2SiC (s) + C(s) + H$_2$O (v) + 5H$_2$. The clusters of SiC nuclei assembled to form nanowire. The nanowires in a preferred orientation grow fast as more SiO vapor and CH₄ gas co-exist in the system. These mechanisms were proposed as oxide-assisted growth for the nanowires growth directly from SiO

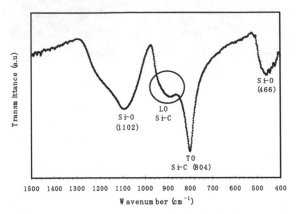

Fig. 3 FT-IR spectrum of core-shell nanowires synthesized at 1350°C for 1 h (SM/O8-H1 specimen).

powder-CH₄ system by Yao et al..[25] Subsequently, side surface of the synthesized SiC nanowires is gradually oxidized to form amorphous SiO₂ outer shell by H₂O vapor, which is a by-product of the formation reaction of SiC nanowires.

3.2 PHOTOLUMINESCENCE PROPERTY OF SYNTHESIZED SiC/SiO₂ NANOWIRES

Fig. 4 shows PL spectrum of SiC/SiO₂ core-shell nanowires synthesized at 1350°C for 1, 3 and 6 h using oxidized Si raw material under 260 nm excitation light at room temperature. Two wide bands of the emission peaks centered about 405 and 470 nm were observed. Chiu et al.[10] have studied the room-temperature PL emission spectrum of the SiCNWs under 250 nm light excitation, and two apparent PL bands were reported to be located at about 390 and 470 nm. These were almost same emission peaks with the present experiment. Moreover, the synthesized nanowires at longer soaking time (3 and 6 h) were measured to compare the emission peaks. Two main board peaks still located at about 405 and 470 nm both from spectrum of specimen synthesized for 3 and 6 h. However, peak at about 405 nm of specimen synthesized for 6 h show very low intensity. The emission peak centered at about 405 nm is attributed to the oxygen discrepancy in the SiO$_x$ amorphous shell layer.[26] The SiO₂ outer layer of nanowires synthesized for 6 h was very thin,[18] resulted in lower PL intensity. Disappearence of SiO₂ layer might be caused from reaction of created-SiO₂ outer layer and Si vapor, because nanowire formation reaction is stopped due to the termination of CH₄ supply. Therefore, intensity of peak at 405 nm of nanowires synthesized for 6 h was very low. The blue emission band centered at about 470 nm originating from the SiCNWs is similar to those reported in other studies.[10, 27] Compared to SiC bulk materials, the emission wavelength for SiCNWs was blue-shifted (from 539 to 470 nm).[28] This might have resulted from the effect of quantum confinement by size reduction or the concentrations of native defects in the SiCNWs.[26, 29]

Many researches have been conducted on the photoluminescence property of SiC nanowires.

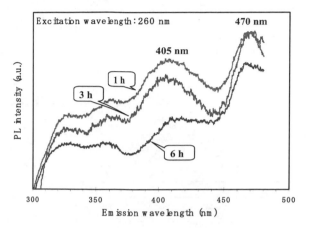

Fig. 4 PL spectrum of SiC/SiO$_2$ core-shell nanowires synthesized at 1350°C for different soaking time of 1, 3 and 6 h using Si raw powder pre-oxidized at 800°C.

Those results indicated that SiC nanowires may be applied as a good light emitting material, since they can emit stable and high-intensity blue-green[10-12] or violet-blue[30] or red[31, 32] light. Due to its blue–green emission property in this study, present SiC/SiO$_2$ nanowires may have an application in blue–green light-emitting diodes (LEDs), and in display devices especially for the environment of high temperature.

3.3 THERMAL CONDUCTIVITY OF ALUMINA CERAMICS CONTAINING SYNTHESIZED NANOWIRES

After sintering at 1350°C for 1 h under Ar atmosphere, physical and thermal properties of the Al$_2$O$_3$ ceramics with and without small amount of nano-sized SiC were measured. Effect of nano-sized SiC addition on bulk density, apparent porosity, mass loss, thermal diffusivity and thermal conductivity of Al$_2$O$_3$ ceramics are summarized in Table 1.

Bulk density of sintered body gradually decreased with an increase of nano-sized SiC. In the case of SiC nanopowder addition, bulk density and porosity of sintered pellet were higher and lower than those of pure alumina pellet (specimen code A), respectively. However, change in bulk density and porosity of Al$_2$O$_3$ with SiC nanowire showed more significant than those Al$_2$O$_3$ with SiC nanopowder. Bulk density (g/cm^3)/porosity (%) of ASp(0.2) and ASw(0.2) were 3.74/1.67 and 3.32/17.04, respectively. Mass loss of all samples showed almost same, average about 1.0 mass%.

Al$_2$O$_3$ disc without any addition occupied thermal conductivity of 39 W/m·K which was identical with previous literature.[33] Thermal conductivity of pellets (included pores) (κ_p) with SiC nanopowder addition presented higher values than those of pellets with SiC nanowire addition at the same added amount. Particularly, ASp(0.035) specimen showed the highest value of 50 W/m·K.

Table 1 Properties of Al₂O₃ samples after sintering at 1350°C for 1 h in Ar atmosphere

Sample code	Bulk density (g/cm³)	Apparent porosity (%)	Mass loss (mass%)	Thermal diffusivity (cm²/s)	Thermal conductivity $(\kappa_p)^*$ (W/m·K)	Thermal conductivity $(\kappa)^{**}$ (W/m·K)
A	3.71	1.76	0.90	0.119	39	40
ASp(0.035)	3.80	0.82	1.14	0.150	50	51
ASp(0.1)	3.75	1.27	1.10	0.136	45	46
ASp(0.2)	3.74	1.67	0.97	0.135	44	46
ASw(0.035)	3.52	10.38	0.94	0.099	31	38
ASw(0.1)	3.45	11.44	0.91	0.093	28	36
ASw(0.2)	3.32	17.04	1.05	0.091	27	38

Remark   *: Thermal conductivity with pores (κ_p) was calculated according to the equation, $\kappa_p = \rho \cdot Cp \cdot \alpha$, where ρ is bulk density, Cp is specific heat and α is thermal diffusivity. (Cp of Al₂O₃ = 0.88 J/g·K)[20]

**: Thermal conductivity without pores (κ) was calculated according to the equation, $\kappa = \kappa_p/[(1-v_d)/(1+v_d)]$, where v_d is porosity of sample.

Thermal conductivity (κ_p) gradually decreased with increasing amount of both SiC nanopowder and SiC nanowire. Thermal conductivity (κ_p) of sintered pellets was related to density of pellets in both systems. In case of thermal conductivity excluded pores (κ), sintered bodies prepared from Al₂O₃-SiC nanowire mixtures possessed almost stable thermal conductivity whereas thermal conductivity of

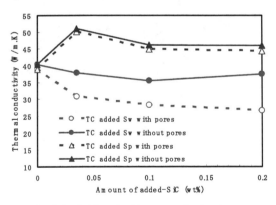

Fig. 5 Thermal conductivity (with and without pores) related to amount of added-SiC into Al₂O₃ samples.

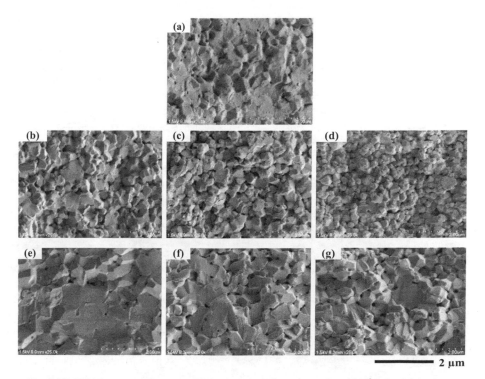

Fig. 6 FE-SEM images of fracture surface of Al$_2$O$_3$ pellets sintered at 1350°C for 1 h in Ar atmosphere: (a) without any addition (A specimen), (b) with SiC nanowires of 0.035 wt% (ASw(0.035) specimen), (c) with SiC nanowires of 0.1 wt% (ASw(0.1) specimen), (d) with SiC nanowires of 0.2wt% (ASw(0.2) specimen), (e) with SiC nanopowder of 0.035 wt% (ASp(0.035) specimen), (f) with SiC nanopowder of 0.1 wt% (ASp(0.1) specimen), and (g) with SiC nanopowder of 0.2wt% (ASp(0.2) specimen)

Al$_2$O$_3$-SiC nanopowder samples decreased with increasing amount of nanosized SiC. The relationship between thermal conductivity with/without pores and amount of added-SiC into Al$_2$O$_3$ samples is shown in Fig. 5.

Phase constitution of the sintered bodies was detected using XRD. All of them showed that all peaks were good matched with that of α-Al$_2$O$_3$. It meant a small amount of added-SiC did not cause a great difference in crystalline phases of the specimens. Nevertheless, change of density seems to be the most significant factor on thermal conductivity. Microstructure of all sintered bodies was observed to confirm these behaviors. Fig. 6 (a) is FE-SEM image of fracture surface of Al$_2$O$_3$ pellet without any addition, and was sintered at 1350°C for 1 h under Ar atmosphere. It composed of Al$_2$O$_3$ grains with average size of about 0.6 μm with small pore distributed at the triple junctions of Al_2O_3 grains.

Fracture surfaces of as-sintered pellets of ASw specimens with SiC nanowire addition of 0.035, 0.1 and 0.2 wt% are shown in Fig. 6 (b)-(d), respectively. Average grain size gradually decreased with increasing amount of SiC nanowire as to be about 0.5, 0.4 and 0.25 μm for addition at 0.035, 0.1 and 0.2 wt%, respectively. Tendency of transgranular fracture increased with increase of nanowire addition. Particularly, at 0.2 wt% addition of SiC nanowire exhibited poor connection of Al_2O_3 grains resulted in the highest porosity than those samples with addition of 0.1 or 0.035 wt%. It was not easy to find the nanowire distribution in samples caused from very small amount of SiC nanowire addition. Fracture surfaces of as-sintered pellets of ASp specimens with SiC nanopowder addition of 0.035, 0.1 and 0.2 wt% are shown in Fig. 6 (e)-(g), respectively as well. In contrast, average grain sizes of ASp specimens were much greater than that of the A specimen. Particularly, ASp(0.035) specimen occupied grain size of about 2.5 μm. However, average grain size decreased with increasing amount of SiC nanopowder at 0.1 and 0.2 wt% (average grain size is 1.8 and 1.0 μm, respectively), whereas pore distribution increased with increasing amount of SiC naopowder.

Apparent from microstructure mentioned above, the thermal conductivity of ASp specimen was higher than those of ASw specimen with the larger average grain size, better connection of Al_2O_3 grains and fewer pore distribution. Change in the thermal conductivity of Al_2O_3 directly reflected by its density in both systems. It was indicated that direct influence of nanowire addition in thermal conductivity of alumina was not significantly observed. The difference in the nanowire addition and nanopowder addition may be caused by the difference in shape, wire and sphere. It is interesting that in the Al_2O_3-nanopowder system, small addition of nanopowder improved density of ceramic and then higher thermal conductivity was achieved. The reason to improve density may be attributed to the presence of SiO_2 on the surface of SiC nanopowder. SiO_2 can be reacted with Al_2O_3 to form lower melting point compounds. Further experimental confirmation is necessary.

4. CONCLUSIONS

The following conclusions were obtained base on the experimental results of this research:

(1) The simple production process and using the low-cost raw material, such as the continuous heating pattern for reaction of coarse silicon powder, make the present approach attractive and outstanding to synthesize SiC/SiO₂ core-shell nanowires. The synthesized nanowires from coarse silicon powder as raw powder under continuous heating pattern possessed core diameter approximately 20-80 nm with wrapped-SiO₂ outer layer of 10-20 in thickness. Length of nanowires was up to 0.5-1 mm.

(2) Two broad photoluminescence peaks located around 405 nm and 470 nm under 260 nm light excitation at room temperature were confirmed.

(3) Sintered bodies prepared from Al_2O_3-SiC nanowire mixtures possessed almost stable thermal conductivity whereas thermal conductivity of Al_2O_3-SiC nanopowder samples decreased with increasing amount of nanosized SiC.

(4) Thermal conductivity of sintered bodies was mostly dominated by density of pellets, both in nanowire-added and nanopowder-added systems.

ACKNOWLEDGEMENT
    This work was supported in part by Global COE Program (Education and Research Center for Material Innovation), MEXT, Japan.

REFERENCES
[1] L. Wang, H. Wada, and L.F. Allard, Synthesis and characterization of SiC whiskers, *J. Mater. Res.*, **7**, 148-163 (1992).

[2] M. Bhatnagar and B.J. Baliga, Comparison of 6H-SiC, 3C-SiC, and Si for power devices, *IEEE Trans. Electron Devices*, **40**, 645-655 (1993).

[3] A. Fisher, B. Schroter, and W. Richter, Low-temperature growth of SiC thin films on Si and 6H–SiC by solid-source molecular beam epitaxy, *Appl. Phys. Lett.*, **66**, 3182-3184 (1995).

[4] N.H. Macmillan, Review: The theoretical strength of solids, *J. Mater. Sci.*, **7**, 239-254 (1972).

[5] E.W. Wong, P.E. Sheehan, and C.M. Lieber, Nanobeam mechanics: elasticity, strength and toughness of nanorods and nanotubes, *Science*, **277**, 1971-1975 (1997).

[6] G. Shen, D. Chen, K. Tang, Y.T. Qian, and S. Zhang, Silicon carbide hollow nanospheres, nanowires and coaxial nanowires, *Chem. Phys. Lett.*, **375**, 177-184 (2003).

[7] M. Zhang, E. Ciocan, Y. Bando, K. Wada, L.L. Cheng, and P. Pirouz, Bright visible photoluminescence from silica nanotube flakes prepared by the sol-gel template method, *Appl. Phys. Lett.*, **80**, 491-493 (2002).

[8] C. Themistos, M. Rajarajan, B.M.A. Rahman, and K.T.V. Grattan, Characterization of silica nanowires for optical sensing, *J. Lightwave Technology*, **27**, 5537-5542 (2009).

[9] Y.B. Li, Y. Bando, D. Golberg, and Y. Uemura, SiO$_2$-sheathed InS nanowires and SiO$_2$ nanotubes, *Appl. Phys. Lett.*, **83**, 3999-4001 (2003).

[10] S.C. Chiu and Y.Y. Li, SiC nanowires in large quantities: Synthesis, band gap characterization, and photoluminescence properties, *J. Cryst. Growth*, **311**, 1036–1041 (2009).

[11] R. Wu, B. Li, M. Gao, J. Chen, Q. Zhu, and Y. Pan, Tuning the morphologies of SiC nanowires via the control of growth temperature, and their photoluminescence properties, *Nanotechnology*, **19**, (2008) 335602 (8pp).

[12] D.F. Feng, T.Q. Jia, X.X. Li, Z.Z. Xu, J. Chen, S.Z. Deng, Z.S. Wu, and N.S. Xu, Catalytic synthesis and photoluminescence of needle-shaped 3C–SiC nanowires, *Solid State Commun.*, **128**, 295-297 (2003).

[13] H. Dai, E. Wang, Y. Lu, S. Fan, and C.M. Lieber, Synthesis and characterization of carbide nanorods, *Nature*, **375**, 769-772 (1995).

[14] X.T. Zhou, N. Wang, H.L. Lai, H.Y. Peng, I. Bello, N.B. Wong, and C.S. Lee, β-SiC nanorods synthesized by hot filament chemical vapor deposition, *Appl. Phys. Lett.*, **74**, 3942-3944 (1999).

[15] S.Z. Deng, Z.S. Wu, Jun Zhou, N.S. Xu, Jian Chen, and Jun Chen, Synthesis of silicon carbide nanowires in a catalyst-assisted process, *Chem. Phys. Lett.*, **356**, 511-514 (2002).

[16] F. Li and G. Wen, A novel method for massive fabrication of β-SiC nanowires, *J. Mater. Sci.*, **42**, 4125-4130 (2007).

[17] W. Khongwong, M. Imai, K. Yoshida, and T. Yano, Synthesis of β-SiC/SiO$_2$ core-shell nanowires by simple thermal evaporation, *J. Ceram. Soc. Jpn.*, **117**, 194-197 (2009).

[18]W. Khongwong, M. Imai, K. Yoshida, and T. Yano, Influence of raw powder size, reaction temperature, and soaking time on synthesis of SiC/SiO₂ coaxial nanowires via simple thermal evaporation, *J. Ceram. Soc. Jpn.,* **117**, 439-444 (2009).

[19]W. Khongwong, K. Yoshida, and T. Yano, Simple approach to fabricate SiC-SiO₂ composite nanowires and their oxidation resistance, *(will be published in Materials Science and Engineering B)*

[20]http://www.accuratus.com/alumox.html (2/2/2010)

[21]H.W. Shim, Y.F. Zhang, and H.C. Huang, Twin formation during SiC nanowire synthesis, *J. Appl. Phys.,* **104**, 063511 (5 pp.) (2008).

[22]Z.J. Li, W.D. Gao. A. Meng, Z.D. Geng, and L. Gao, Large-scale synthesis and Raman and photoluminescence properties of single crystalline β-SiC nanowires periodically wrapped by amorphous SiO₂ nanospheres 2, *J. Phys. Chem. C,* **113**, 91-96 (2009).

[23]S. Dhage, H.C. Lee, M.S. Hassan, M.S. Akhtar, C.Y. Kim, J.M. Sohn, K.J. Kim, H.S. Shin, and O.B. Yang, Formation of SiC nanowhiskers by carbothermic reduction of silica with activated carbon, *Mater. Lett.,* **63**, 174-176 (2009).

[24]G.W. Meng, L.D. Zhang, Y. Qin, C.M. Mo, and F. Phillipp, Synthesis of β-SiC nanowires with SiO₂ wrappers, *Nanostruct. Mater.,* **12**, 1003-1006 (1999).

[25]Y. Yao, S.T. Lee, and F.H. Li, Direct synthesis of 2H-SiC nanowhiskers, *Chem. Phys. Lett.,* **381**, 628-633 (2003).

[26]H.K. Seong, H.J. Choi, S.K. Lee, J.I. Lee, and D.J. Choi, Optical and electrical transport properties in silicon carbide nanowires, *Appl. Phys. Lett.,* **85**, 1256-1258 (2004).

[27]X.J. Wang, J.F. Tian, L.H. Bao, C. Hui, T.Z. Yang, C.M. Shen, and H.J. Gao, Large scale SiC/SiOx nanocables: Synthesis, photoluminescence, and field emission properties, *J. Appl. Phys.,* **102**, 014309 (6pp) (2007).

[28]G. Xi, Y. Liu, X. Liu, X. Wang, and Y. Qian, Mg-catalyzed autoclave synthesis of aligned silicon carbide nanostructures, *J. Phys. Chem. B,* **110**, 14172-14178 (2006).

[29]J.J. Niu and J.N. Wang, A simple route to synthesize scales of aligned single-crystalline SiC nanowires arrays with very small diameter and optical properties, *J. Phys. Chem. B,* **111**, 4368-4373 (2007).

[30]X.M. Liu and K.F. Yao, Large-scale synthesis and photoluminescence properties of SiC/SiOx nanocables, *Nanotechnology,* **16**, 2932-2935 (2005).

[31]K.F. Cai, A.X. Zhang, J.L. Yin, H.F. Wang and X.H. Yuan, Preparation, characterization and photoluminescence properties of ultra long SiC/SiOx nanocables, *Appl. Phys. A,* **91**, 579-584 (2008).

[32]F. Gao, W. Yang, H. Wang, Y. Fan, Z. Xie and L. An, Controlled Al-doped single crystalline 6H-SiC nanowires, *Crystal Growth & Design,* **8**, 1461-1464 (2008).

[33]R. Morrell, Handbook of Properties of Technical and Engineering Ceramics: Part 1 An Introduction for the Engineer and Designer, HMSO, London.

FABRICATION AND CHARACTERIZATION OF MULTIFUNCTIONAL ZnO-POLYMER NANOCOMPOSITES

Hongbin cheng, Qian Chen, Qing-Ming Wang
Department of Mechanical Engineering and Materials Science
University of Pittsburgh
Pittsburgh, PA, USA

ABSTRACT

In this paper, we present our recent study on the nanocomposite in which the polyimide was used as the polymeric matrix and the vertically grown ZnO nanowires as the inclusions such that an anisotropic nanocomposite is formed by in-situ polymerization. The top electrode was deposited using a shadow mask and DC magnetron sputtering for electrical property characterization of the material. Scanning electron microscopy (SEM) examination indicates a uniform distribution of ZnO nanowires in polyimide matrix. The frequency spectra of dielectric permittivity and loss tangent, and electrical I-V curve of the nanocomposites were characterized. The results indicate that the dielectric permittivity is significantly enhanced due to the addition of ZnO nanowires.

INTRODUCTION

Polymer-based nanocomposites are becoming an attractive set of materials due to their multifunctional properties and many potential applications [1]. These materials are expected to possess unique electric, magnetic, optical, and mechanical properties which can be significantly different from those of single component materials. Polyimides are a class of polymers with material properties that are used in many applications. Polyimides are usually synthesized by reaction of a dianhydride and a diamine [2], and show stability at high temperatures and exhibit high glass transition temperatures in the range of 300-400°C. These materials offer mechanical properties, as well as chemical resistance and low dielectric constant [3]. Due to their flexibility and resistance to heat and chemicals, polyimides have been used in the electronics industry for flexible cables, as an insulating film. Polyimide is also used as a high-temperature adhesive, as a mechanical stress buffer, and as passivation layer in the manufacture of digital semiconductor and MEMS chips. Some polyimides can be used as photoresists in the microfabrication processing.

To date, one-dimensional ZnO micro/nanostructures have been attracting much attention for wide potential applications due to their unique electrical, piezoelectric, optoelectronic, and photochemical properties [4-6]. Therefore, in this paper, we present our recent study on the nanocomposite in which the polyimide was used as the polymeric matrix and vertically aligned ZnO nanowires as the inclusions was grown on the silicon substrate through a simple hydrothermal route[7]. Scanning electron microscopy (SEM) examination indicates a uniform distribution of ZnO nanowires in polyimide matrix. Effective dielectric constant and conductivity of the nanocomposites is enhanced due to the addition of ZnO nanowires.

FABRICATION OF ZnO-POLYIMIDE NANOCOMPOSITES

For polymeric matrix composites, the assembly of composite materials are the key to success. ZnO nanowires, however, have rarely been used as inclusions in a polymer matrix because of the difficulty in achieving uniform dispersion. Intrinsic van der Waals attraction among nanowires, in combination with their high surface area and high aspect ratio, often leads to significant agglomeration, thus preventing efficient transfer of their superior properties to the matrix [8]. Here, we developed a self-alignment method to efficiently disperse ZnO nanowires into a given polyimide matrix on a nanoscale level to produce the novel ZnO-polyimide nanocomposite.

The self-alignment method is a kind of bottom-up fabrication process. First, High-quality vertical self-aligned ZnO nanowire arrays were synthesized on the top of substrates with Au electrode according to a two-step hydrothermal method. The polymer used as a matrix was low stress commercial polyimide PI-2611 bought from HD Microsystems Inc. Since the pure polyimide was too thick to completely cast into the ZnO nanowires layer, it was first diluted by the solvent T9039 (HD Microsystems Inc.) with weight ratio 1:1. The mixed solution was prepared by homogenizing for 1 hour (350 rpm with a 6 mm diameter rotor) and stood at room temperature overnight for stabilization of the solution. The homogenous solution was dropped onto the surface of the substrate covered by ZnO nanowires layer, then the sample was placed into the vacuum oven (Isotemp 282A) and held under vacuum for one minute to allow all bubbles to dissipate out solution. After that, the sample was taken out and kept in the normal pressure for 30 minutes to make sure the polyimide flow in the nanowires layer as far as possible and relax. Two steps of spin coating process were employed in the deposition using G3P-8 specialty coating system (Cookson Electronics Inc.): the first run was 500 rpm for 5 seconds to make the polyimide to gradually cover the substrate and the second run was a high speed 2000 rpm for 30 seconds according to the desired thickness of films. A soft-bake was done on the hot plate at 120 °C for 2 minutes after coating. Subsequently, the soft-baked films were cured at 350 °C for 30 minutes in the muffle oven to convert the polyamic acid into a fully aromatic, insoluble polyimide to obtain the solvent-free ZnO-polyimide nanocomposite. Figure 1 shows the morphology of polyimide-ZnO nanowire nanocomposites. The homogenous dispersion of ZnO nanowires in a polymer matrix was achieved, and the thickness of nanocomposite film was 3.2 um with a flat top surface. The surface roughness of the ZnO-polyimide nanocomposite was evaluated by Scanning Probe Microscope (SPM) (Veeco, Dimension V). Figure 2 shows scanning results. The surface of the nanacomposite film is quite smooth and the variation of the height is less than 100 nm.

CHARACTERIZATION OF OF ZnO-POLYIMIDE NANOCOMPOSITES

As we know that for a common two-plate capacitor, the capacitance (Cp) could be written as:

$$C_p = \frac{\varepsilon \, \varepsilon_0 A}{d} \tag{1}$$

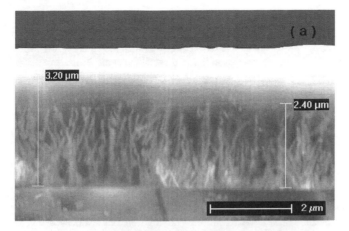

Figure 1. SEM image of ZnO-Polyimide nanocomposites

Figure 2. SPM image of top surface of the nanocomposite film

where ε is the dielectric constant of the nanocomposite, A is electrode area, d is the thickness of the sample, and ε_0 is the permittivity of free space where $\varepsilon_0 = 8.854 \times 10^{-12}$ F/m. Since ZnO nanocomposites prepared by self-algnment method have a similar two parallel plate structure to the normal capacitor. So we can rewrite equation (1) to get the dielelctric constant of the nanocomposite ε like below:

$$\varepsilon = \frac{C_p d}{\varepsilon_0 A} \qquad (2)$$

The dielectric loss is given by

$$\varepsilon' = D \cdot \varepsilon \qquad (3)$$

D is the dissipation factor.

The dielectric capacitance and the loss tangent of the nanocomposite devices were measured at frequencies of 100Hz, 1 kHz, 10 kHz, 100 kHz, and 1 MHz, respectively, using an Agilent precision LCR Meter (model 4284A, Agilent Technologies). The dielectric permittivity and loss versus frequency are plotted in figure 3. Addition of ZnO nanowires to the polymer has a dramatic effect on the dielectric constant and loss. At 1 KHz the dielectric constant of the nanocomposite increases by nearly 3 times of the pure polyimide due to the interfacial effect between ZnO nanowires and the polymer matrix. Meanwhile, it is noted that the dielectric permittivity decrease as the frequency increases from 100 Hz to 1MHz. For the tangent loss, the nanocomposite is also much higher than the pure polymer and significantly changes in the low frequency range, which has the maximum magnitude at 100Hz. When the frequency is more than 100 kHz, it becomes stable again. The improvement of the dielectric constant can be attributed to the interface polarization induced between ZnO nanowires and polyimide matrix, which lead to the increase of the dielectric constant capacitance and dielectric loss of the composites.

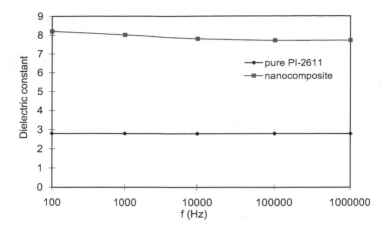

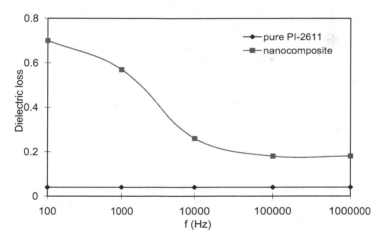

Figure 3. Dielectric constant and loss as a function of frequency

The room temperature IV characteristic of the device was measured by a Keithley 238 high current source measure unit (Keithley Instrument Inc.). The result in Figure 4 shows that the pure polyimide film device is insulated. When inclusions of ZnO nanowires were added into the polymer, the conductivity of the nanocomposite device was significantly increased compared with the pure polymer film. Though polyimide deposits on top of the ZnO nanowires, some of the nanowires are

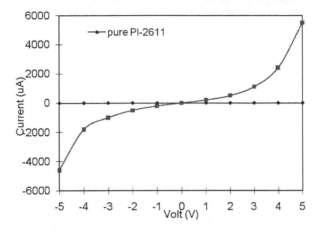

Figure 4. IV characteristic of ZnO nanocomposites

actually not buried under the polyimide and have direct connection with top electrode; thus the nanocomposite overall shows semiconductor I-V behavior.

CONCLUSION

In summary, the thin film nanocomposite with uniformly dispersed ZnO nanowires in the polyimide matrix was achieved using the self-alignment method. This fabrication process included the controlled growth of ZnO nanowires on the substrate and the polymer spin-coating. The resultant nanocomposite devices exhibited great dielectric constant and conductivity enhancement at room temperature due to the interfacial effect between ZnO nanowires and the polymer matrix. These nanocomposites combing the superb properties of ZnO nanowires with the polyimide polymer matrix provide a smart material candidate for multifunctional applications that require self-sensing and self-actuation capabilities. The self-alignment method provides a bright route to combine superb properties of nanomaterials with the lightweight, flexibility, and manufacturability of dielectric polymers for future generations of multifunctional materials. The novel nanocomposites prepared by this technique may have tremendous impact in intelligent materials and structure applications, including piezoelectric sensors and actuators, biological sensors, structural health monitoring and vibration control in numerous industrial, civil, medical and aerospace applications.

REFERENCES

[1] C. Sanchez, B. Julian, P. Belleville, and M. Popall, Applications of hybrid organic–inorganic nanocomposites, J. Mater. Chem., **15**, 3559 (2005).

[2] K. Saunders, Organic Polymer Chemistry, Chapman & Hall (1988).

[3] K. Jordan, X. Li, O. Iroh, , SPE ANTEC , 2436-2439 (1999).

[4] Q. Wan, Q. H. Li, Y. J. Chen, T. H. Wang, X. L. He, J. P. Li, and C. L. Lin, Fabrication and ethanol sensing characteristics of ZnO nanowire gas sensors, Appl. Phys. Lett., **84**, 3654 (2004).

[5] K. Hara, T. Horiguchi, T. Kinoshita, K. Sayama, H. Sugihara, and H.Arakawa, Sol. Energy Mater. Sol. Cells, **64**, 115 (2000).

[6] Y. B. Li, Y. Bando, and D. Golberg, ZnO nanoneedles with tip surface perturbations: Excellent field emitters, Appl. Phys. Lett. **84**, 3603 (2004).

[7] L. Greene, M. Law, D. Tan, M. Montano, J. Goldberger, G. Somorjai, and P. Yang, General route to vertical ZnO nanowire arrays using textured ZnO seeds, NANO letters, **5**, 1231-1236 (2005).

[8] Z. Ounaies, C. Park, K.E. Wise, E.J. Siochi, J.S. Harrison, Electrical properties of single wall carbon nanotube reinforced polyimide composites, Composites Science and Technology **63**, 1637–1646 (2003).

HYBRID NANOSTRUCTURED ORGANIC/INORGANIC PHOTOVOLTAIC CELLS

S. Antohe[1], I. Enculescu[2], Cristina Besleaga[1], Iulia Arghir[1], V. A. Antohe[1]
V. Covlea[1], A. Radu[1], L. Ion[1]

[1]University of Bucharest, Faculty of Physics, P.O. Box MG-11, Magurele, Ilfov, 077125, ROMANIA
[2]National Institute for Materials Physics, P.O. Box MG-7, Magurele, Ilfov, 077125, ROMANIA
*Corresponding author: Tel.:+ 40214574535, E-mail: santohe@solid.fizica.unibuc.ro

ABSTRACT
    In the last decade, the hybrid structures based on nanostructured inorganic materials and organic thin films seem to be promising prototypes for producing low-cost photovoltaic cells.
    The absorption of the organic materials, in the visible range of the solar spectrum is strong, but based on an excitonic mechanism. Taking into account that the diffusion length of the excitons in organic semiconductors is of 30-80 nm, one way to improve the extraction of the charge carriers, will consist in significantly increasing the area of the interface between Donor (D) and Acceptor (A) components of the heterostructure, then reducing the dimensions of D/A heterojunctions to the dimensions of exciton diffusion length in the organic absorber.
    Our results on producing and characterizing inorganic/organic photovoltaic structures based on ZnO nanostructured electrodes will be reported. Two types of structures were investigated: i) ZnO wire arrays electrodes photosensitized with copper phthalocyanine (CuPc) as organic absorber and ii) nanostructured ZnO electrodes photosensitized with CuPc.
    The hybrid cells, based on the nanostructured ZnO electrode/CuPc were produced and characterized by structural, morphological, electrical and optical measurements. Promising photovoltaic response was obtained with an EQE strongly dependent on the morphology of the inorganic electrode and the design of the cells, but a lot of work is yet to be performed to improve their performances as photo elements.

INTRODUCTION

    The study of the materials and designs of different types of photovoltaic structures represents a priority research field. Systematic research must be done on each layer of these devices, on the interaction between successive layers and on improving the transparent conducting oxide (TCO) contacts. Such contacts have to be made from low-cost materials, they must have high conductivities and high transparencies in the NIR-VIS range of the solar spectrum. So far, indium tin oxide (ITO) is the most representative TCO material. ITO films with resistivities as low as 210^{-4} $\Omega \cdot$cm and high optical transparency in the visible region of the spectrum were successfully produced [1,2]. However, ITO includes expensive In_2O_3 and requires high substrate temperatures for obtaining low resistivities. Also ITO is reported to be structurally and electrically unstable at ~150 °C [3].
    Zinc oxide has received much attention, due to its remarkable properties and to the diversity of its technological applications, including transparent conducting electrodes of solar cells. ZnO is a direct band-gap semiconductor, exhibiting normally n-type conduction. Its large band-gap and a very good flexibility in impurity doping, while still retaining reasonably high carrier mobilities, recommends it as a good candidate for use as TCO [4].
    In the last decade, hybrid structures based on nanostructured inorganic materials and thin films of organic semiconductors have attracted a great deal of interest among scientists involved in the research efforts aiming at producing low-cost solar cells. Using of organic semiconductors is expected to result in the desired reduction of the costs. Among the organic semiconductors envisaged to be used in such structures, metal-doped phthalocyanines (MePc, with Me=Cu, Mg, Zn, etc.) are the most studied, due

to their peculiar optical properties. Their optical absorption in the visible range of the solar spectrum is strong, but based on an excitonic mechanism. Most of the photogenerated excitons annihilate by direct recombination before the occurring of charge separation in the internal field of the structure. This charge extraction problem can be avoided by creating a large area heterostructure at the interface with an inorganic semiconductor. Such an approach should also reduce the series resistance of the structure. Excitons photogenerated near this interface will dissociate by electronic transfer to the inorganic semiconductor. Typical values of the diffusion length of the excitons in organic semiconductors are in the range 30-80 nm, while in order to achieve the required efficiency in light absorption, the absorber layer has to be at least 100 nm thick. Therefore, one way to improve the extraction of the charge carriers consists in increasing significantly the area of the interface between the two components of the heterostructure.

Up to now, a number of ZnO nano or micro materials with different morphologies and interesting structures, such as nanowires , nanobelts [5] , nanocombs [6], nanorings [7], and nano/micro tubes [8], have been successfully synthesized with different methods. Compared with the above mentioned morphologies, hollow tubular structures have a larger surface area.

Several methods of synthesizing arrays of tubular nano or micrometer ZnO have been employed, including thermal evaporation [9], plasma-assisted molecular beam epitaxy using a two-step process [10], hydrothermal method [11], or vesicle–template fusion [12].

In this paper we report on the preparation and physical properties of three types of nanostructured ZnO films, suited for use as photosensitized electrodes in hybrid organic/inorganic photovoltaic cells. Details about fabrication and experimental methods used for the characterization of the obtained structures are presented.

Hybrid structures of type ITO/(nc)ZnO/CuPc/Al have been produced and characterized, and a comparative discussion of the experimental results on the efficiency of the cells based on such types of films is given. It was found that the cells based on nanocrystalline tubular structures had the best performance in the series we have produced.

EXPERIMENTAL PROCEDURES

Fabrication of ZnO wire arrays

ZnO wires arrays were electrochemically deposited by using a template method, with ion track polymer membranes used as templates. Polycarbonate foils (Makrofol N, Bayer), 30 μm thick, were irradiated with swift heavy ions (e.g. U with specific energy 11.4 MeV/nucleon) at fluencies in the range 10^4-10^8 ions/cm^2. Details about preparing the template are given elsewhere [13]. The electrochemical deposition was performed using a VoltaLab potentiostat controlled by a computer. ZnO was deposited from a 0.1 M $Zn(NO_3)_2$ aqueous solution at 90°C. A three-electrode configuration was used with a platinum counter-electrode and a commercial saturated electrode as reference. The deposition potential was chosen at -800 mV vs. (SCE).

Fabrication of Nanostructured ZnO Films

Thin films consisting of ZnO hollow hexagonal columns were prepared by electrochemical deposition using as substrate optical glass covered with a 90 nm ITO thick film.

A standard three electrode configuration was used, with a platinum counter-electrode, a commercial saturated calomel electrode (SCE) as reference and ITO as working electrode. ZnO was deposited from 0.1M Zn (NO) 3 aqueous solutions, at 90°C, using a double wall thermostatic cell. The applied voltage was swept in the range from -400 mV to -1100 mV, with a rate of 0.3 mV/s. Three sweeps were performed for the deposition of a layer.

For comparison purposes, thin ZnO films were also produced by pulsed laser deposition (PLD) technique, using an excimer laser source KrF* (λ=248 nm, τ_{FWHM}=25ns) for ablation of In-doped ZnO targets. Details regarding their structure and optical properties were given elsewhere [14]. Table 1 summarizes the growth conditions and some of the parameters characterizing the films.

Table 1: PLD deposition parameters for ZnO thin films

Sample	Target	Substrate	Tempe-rature [°C]	Pressure [mbar]	Flu-ency	Thickness [nm]	Rough-ness [nm]
ZnO-1	In$_2$O$_3$:ZnO 20%	glass	500	5	1.6	513	7
ZnO-2	In$_2$O$_3$:ZnO 3%	glass	500	5	1.6	430	3

Fabrication of ZnO/CuPc photovoltaic structures

As a second step, after the preparation of ZnO wire arrays or ZnO nanostructured thin films, an organic dye, copper phthalocyanine (CuPc), was deposited by vacuum sublimation, on the ZnO thin films or wire arrays. The organic films were deposited at 437 °C and the work pressure was 3×10^{-5} Torr. An aluminum top contact was deposited by vacuum thermal evaporation to complete the photovoltaic structures. A cross section of the prepared structures is shown in Figure 1.

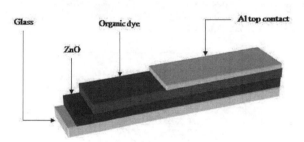

Figure 1: Schematic image of a photovoltaic ZnO/CuPc device.

Characterization Techniques

The morphological features of the samples were evidenced by Atomic Force Microscopy (AFM) and Scanning Electron Microscopy (SEM). The crystallinity of the thin films was characterized by X-ray diffraction (XRD) at low angle incidence with a detector scan method, using a Bruker D8 Discover diffractometer. XRD spectra were recorded by using Cu-K$_{\alpha 1}$ line, λ=1.5406 Å.

Transmission spectra were recorded at room temperature using a UV-VIS Perkin-Elmer Lambda 35 spectrophotometer. Action spectra were performed with a set-up consisting of a Cornerstone 130 monochromator, a Keithley 2400 SourceMeter and a Keithley 6517a electrometer, controlled by a computer. The current-voltage (I-V) characteristics of the cells under illumination with monochromatic light, at wavelengths corresponding to the maximum in action spectrum for each sample, were measured at room temperature.

RESULTS AND DISCUSSIONS

ZnO Nanowires Arrays

Figure 2 shows a general view of the homogeneous array of wires grown in the above-mentioned conditions, as revealed by a scanning electron microscope.

Figure 2. SEM micrograph of ZnO wires grown at 90°C.

XRD studies have been carried out on ZnO wire arrays, by using a detector scan method, with incident X-ray beam at 2.5° incidence (Figure 3). The wires exhibit a polycrystalline structure; the hexagonal (wurtzite) ZnO phase is present.

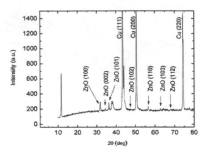

Figure 3. XRD pattern of ZnO wire array, showing a wurtzite-type crystalline structure.

Considering the integral breadths, the peaks are relatively large, providing evidence of the presence of nanostructured grains in the structure of the wires. Using the Scherrer formula, a crystalline coherent zone of 23 nm for (100) direction was obtained. The peaks at 2Θ =11.22 and 37.82 might be ascribed to simonkolleite compound ($Zn_5(OH)_8Cl_2 \cdot H_2O$, see PDF2 77-2311). In the case of this sample, KCl was added to the electrochemical bath, to improve its conductivity. At the chosen value of the deposition potential, the formation of $Zn_5(OH)_8Cl_2 \cdot H_2O$ was not suppressed, probably due to a too high chloride ion concentration [4].

Nanostructured ZnO Thin Films Grown on ITO Covered Optical Glass

Figures 4.a and 4.b shows SEM images of a ZnO thin film grown on ITO covered optical glass. It consists of columns showing a perfect hexagonally faceted morphology, with an edge of about 500-700 nm, and uniform wall thickness of about 70 nm.

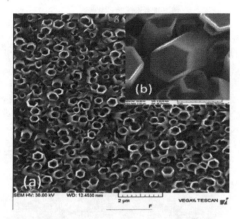

Figure 4. (a) SEM image of a ZnO film grown on ITO covered optical glass substrate, showing hexagonal columns. (b) A detailed view of a ZnO hexagonal column.

XRD pattern recorded in the case of this sample reveals a well formed hexagonal wurtzite structure. The intensity of the (002) diffraction is much higher than that of other peaks. All the observed peaks are broadened as a consequence of the nanostructuring of the film, also seen in SEM micrographs in Figure 5. Using the Scherrer formula a crystalline coherent zone of 10 nm for (100) direction was obtained.

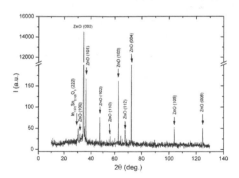

Figure 5. XRD pattern of the ZnO film with a hexagonal-columnar morphology.

ZnO Thin Films Deposited by PLD

Figures 6 and 7 show the surface morphology of PLD deposited thin films, obtained by AFM scanning in noncontact mode. Scanned area (5 μm) shows elongated grains, oriented normally to the substrate. A small difference between roughness values of ZnO-1 and ZnO-2 was observed (see Table 1).

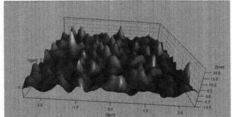

Figure 6. AFM topographic scan of ZnO-1 thin film.

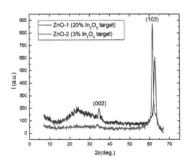

Figure 7. AFM topographic scan of ZnO-2 thin film.

Structural analysis of the samples by X-ray diffraction, revealed that the films consist of a hexagonal-close-packed wurtzite type phase ZnO, (103) preferentially oriented in the growth direction. No In_2O_3 crystalline phase was found, within the detection limit (Figure 8).

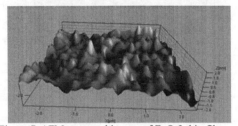

Figure 8. X-ray diffraction pattern of ZnO films: black line corresponds to ZnO-1 sample, red line corresponds to ZnO-2.

Optical transmission spectra, measured in the 300 nm-1100 nm range, show that the average optical transmission in the VIS range is better than 70%, Figure 9.

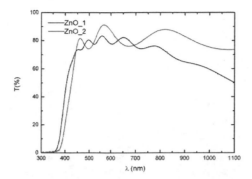

Figure 9. Transmission spectra of PLD deposited ZnO films, recorded at room temperature.

The thickness of the films, indicated in Table 1, was determined from optical transmission spectra, with the method described in [6]. The temperature dependence of the electrical resistivity of the two investigated samples is shown in Figure 10.

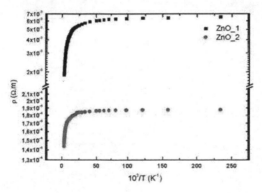

Figure 10. Temperature dependence of the resistivity of the PLD deposited ZnO samples.

The resistivity only changes by a factor of 3 in the case of ZnO_1 sample (1.2 in the case of ZnO_2 sample) over a large temperature range, from 300 K down to 4 K. For both cases, $\rho(T)$ dependence is more pronounced in the high temperature range, above 100 K. The Hall coefficient R_H is practically constant over the whole investigated temperature range, except for temperatures near room temperature where it is slightly temperature dependent (Figure 11). Note that the sample ZnO_1 shows a p-type conduction, while ZnO_2 is n-type.

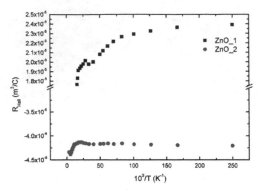

Figure 11: Temperature dependence of Hall coefficient, obtained in the case of PLD deposited ZnO sample.

Assuming that the Hall scattering factor has a value of unity, a free carrier density of 5.3×10^{25} m^{-3} was obtained for the sample ZnO_1 at room temperature (1.4×10^{26} m^{-3} for the sample ZnO_2).

The Hall mobility is also slightly temperature dependent, following a power law $\mu \sim T^r$ at temperatures above 100 K, with r = 0.6 for ZnO_1, r = -0.2 for the ZnO_2 sample, respectively, as can be seen in Figure 12.

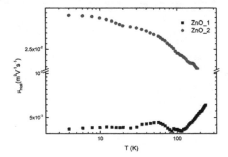

Figure 12. Temperature dependence of the Hall mobility obtained in the case of PLD deposited ZnO samples.

The interesting fact was that different conduction types (n or p) were obtained by varying the chemical composition of the targets. Consequently, one can envisage the use of such films in hybrid structures with both p-type and n-type organic semiconductors. On the other hand, larger charge carriers mobilities results in reduced series resistance of the cells, hence in a better efficiency.

Photovoltaic Characterizations of the Hybrid Structures ZnO/CuPc

The external quantum efficiency (EQE) of a solar cell gives information on the current that a given cell will produce when illuminated by a particular wavelength; it measures the fraction of incident photons for an electron-hole pair collected at its electrodes, in short-circuit conditions. It does not consider reflected or transmitted photons.

$$EQE = \frac{I}{e \cdot A_p \cdot \varphi}$$
(1)

where I is the short-circuit current due to incident photons with λ wavelength, e is the elementary charge, Φ is the incident photons flux, and A_p is the illuminated area of the structure.

Action Spectra of Au/ZnO Wire Array/CuPc/ZnO Structures and Absorption Spectra of CuPc Thin Films.

The spectral dependence of external quantum efficiency (EQE) of photovoltaic structure, Au/ ZnO wires array/CuPc/Al measured at ambient temperature (25°C) is shown in Figure 13.

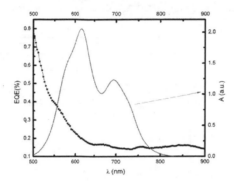

Figure 13. EQE characteristics (black) and absorption spectrum (red) for a ZnO/CuPc/Al photovoltaic cell based on ZnO wires array.

Under illumination with higher energy photons (wavelengths less than 650 nm, Fig. 13), EQE increases abruptly with a maximum of 0.8, while its value does not exceed 0.2 at wavelengths larger than 650 nm.

Action spectra of Al/PLD ZnO/CuPc/Al structures and Absorption spectra of CuPc thin films.

The action spectra of ZnO/CuPc/Al photovoltaic cell together with absorption spectrum of the same structure are shown in Figure 14, revealing a batic behavior. The maximum in the action spectrum (at 660 nm) is slightly red-shifted as compared to the local maximum at 610 nm in the optical absorption.

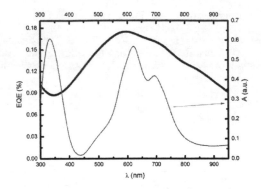

Figure 14. Action spectrum (black) and absorption spectrum (red) for a ZnO/CuPc/Al photovoltaic cell.

A maximum value of 0.18 was obtained for the EQE. The fourth quadrant of I-V characteristics of the cell, measured at illumination under AM 1.5 conditions, is shown in Figure 15.

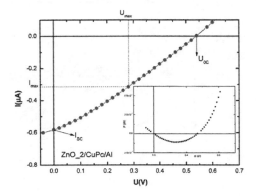

Figure 15. Current-voltage characteristics for ZnO/CuPc/Al photovoltaic cell in the fourth quadrant under AM 1.5 illumination.

Action spectra of ITO/ZnO hollow hexagonal column array/CuPc/Al structures and Absorption spectra of CuPc thin film.

The optical absorption spectrum of CuPc film and the action-spectrum of ITO/ZnO hollow hexagonal column arrays/CuPc/Al structures are plotted in Figure 16.

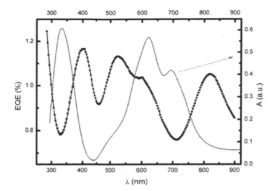

Figure 16. EQE characteristics (black) of ITO/ZnO hexagonal column array/CuPc/Al structures and absorption spectrum (red) of a CuPc layer.

One can notice the anti-batic response of the ZnO/CuPc layer, i.e., a local maximum in photocurrent is obtained at photon wavelengths where the absorption is about at its minimum.

The values of EQE are significantly increased for this structure, as compared to the values measured in the case of structures based on ZnO obtained by PDL or ZnO wire arrays.

A possible explanation for this significant growth of the external quantum efficiency is that the heterojunction area of the structured cell is larger than that of the flat cell. A higher active surface means higher charge collection efficiency at the ITO/ZnO electrode: the excitons photogenerated in the region adjacent to the interface, within at most an exciton diffusion length, can reach the interface with the inorganic electrode, where they can dissociate.

CONCLUSIONS

Nanostructured ZnO for applications in the field of hybrid organic/inorganic photovoltaic cells was produced by different techniques. Structural, morphological and photovoltaic characterizations of the obtained structures were performed.

The external quantum efficiencies of the prepared photovoltaic structures were determined and their performances related to the properties of the nanostructured ZnO component films are discussed.

EQE value was significantly increased in the case of ITO/ZnO hexagonal column array/CuPc/Al structure, as compared to the values measured in the case of structures based on ZnO obtained by PLD or ZnO wires array electrodes. We infer that the observed trend is related to a significant increase of the active area at the CuPc/(nc)ZnO interface. A parallel way of improving the performance of such type of cells consists in increasing the exciton diffusion length, by improving the crystalline structure of the organic layer. Work is currently in progress in this direction.

ACKNOWLEDGEMENTS

The authors acknowledge the financial support of National Authority for Scientific Research of Romania, under grant no. 11-060.

REFERENCES

[1] I. Hamberg, C.G. Granqvist, *J. Appl. Phys.*, **60,** R123 (1986),

[2] Y. Shigesato, S. Takaki, T. Haranoh, *J. Appl. Phys.*, **71**, 3356 (1992).

[3] D.C. Paine, E. Chason, E. Chen, D. Sparacin, H.-Y. Yemon, *Mat. Res. Soc. Symp. Proc.*, **623**, 245, (2000).

[4] Toshio Kamiya, Masashi Kawasaki, *MRS Bull.,* **33**, 1061 (2008).

[5] L.L. Zhang, C.X. Guo, J.G. Chen, J.T. Hu, Chin. Phys. 14 (2005) 586

[6] U. Ozgur, Y.I. Alivov, C. Liu, et al. J. Appl. Phys. 98 (2005) 041301

[7] X.Y. Kong, Y. Ding, R.S. Yang, Z.L. Wang, Science 303 (2004) 1348

[8] W.Z. Xu, Z.Z. Ye, D.W. Ma, H.M. Lu, L.P. Zhu, B.H. Zhao, Appl. Phys.Lett. 87 (2005) 093110

[9] Tianjun Sun, Jieshan Qiu, Materials Letters 62 (2008) 1528 – 1531

[10] Yan Jian-Feng, Lu You-Ming, Liang Hong-Wei, Liu Yi-Chun, Li Bing-Hui, Fan Xi-Wu, Zhou Jun-Ming, Journal of Crystal Growth 280 (2005) 206–211

[11] J.H. Yang, J.H. Zheng, H.J. Zhai, L.L. Yang, Y.J. Zhang, J.H. Lang, M. Gao, Journal of Alloys and Compounds 475 (2009) 741–744

[12] Yanyan Ding, Zhou Gui, Jixin Zhu, Shanshan Yan, Jian Liu,Yuan Hu, Zhengzhou Wang, Materials Letters 61 (2007) 2195 – 2199

[13] C. Tazlaoanu, L. Ion, I. Enculescu, M. Sima, Elena Matei, R. Neumann, Rosemary Bazavan, D. Bazavan, S. Antohe, *Physica E: Low-dimensional Systems and Nanostructures*, **40**, 2504 (2007).

[14] R. Bazavan, L. Ion, G. Socol, I. Enculescu, D. Bazavan, C. Tazlaoanu, A. Lorinczi, I. N. Mihailescu, M. Popescu, S. Antohe, *J. Optoel. Adv. Mater.*, **11**, 425 (2009).

ENHANCED PHOTOVOLTAIC EFFECT USING NANOSTRUCTURED MULTI-LAYERED PHOTOELECTRODE

M. Ramrakhiani and J.K. Dongre

Dept. of Postgraduate Studies and Research in Physics & Elecronics,
Rani Durgawati University, Jabalpur. 482001, India.

ABSTRACT

Higher conversion efficiencies can be achieved by using layers of different band gaps in photoelectrode (PE) of photoelectrochemical (PEC) solar cells. Such cells have been prepared with PE consisting of 3 layers CdS having different band gaps. The band gap of CdS is tuned by the size and structure of nano-CdS obtained by changing the preparation parameters. The PE was prepared by depositing three layers of CdS having different sizes and morphologies onto titanium substrate by chemical bath deposition. Scanning electron micrographs show honeycomb morphology for the 1st layer of CdS, where as flower-like and nanowire morphology is exhibited by 2nd and 3rd layers respectively. Optical absorption spectra of the films were recorded and band gap values were computed as 2.2, 2.4 and 2.48 for 1st, 2nd and 3rd layer respectively. PEC cell was fabricated using a two electrode configuration, comprising n-CdS thin films on titanium as PE and graphite as counter electrode. Polysulfide was used as redox electrolyte. Enhanced photovoltaic effect is observed in PEC solar cells by using the multi-layered PE as compared to any one single CdS layer as PE. For multi-layered PE, efficiency is about 10 times as compared to single honeycomb structured PE.

1. INTRODUCTION

For photovoltaic (PV) technology to be a serious contender as an alternative energy source, the electricity generated must be reasonably cost-effective compared with current fossil fuel sources. This requires that the efficiency of the PV cells be increased and their manufacturing costs be reduced. While photovoltaic cells based on bulk semiconductors can provide quite high efficiency, their high manufacturing cost currently limits their use. Nanoparticles (NPs), nanorods (NRs) and nanowires (NWs) have been used to improve charge collection efficiency in solar cells to demonstrate carrier multiplication and to enable low-temperature processing of photovoltaic devices [1]. The development of thin film solar cells makes use of at least two kinds of semiconducting layers: a wide bandgap window material and a narrow bandgap absorber material. The two most researched absorber materials are CdTe [2] and Cu(In,Ga)Se$_2$ [3] and in both cases the majority of researchers used a n-CdS layer as a window material [4-6]. In this multi-junction configuration transmission and thermalization losses of hot carriers can be reduced. But often it is difficult to grow the desired materials on top of each others due to mismatched lattice constants. Furthermore, the semiconductor materials have to be of high quality, which is usually achieved for material with similar lattice constant. These two boundary conditions reduce the number of candidates for a high efficiency multijunction solar cell [7]. These boundary conditions are suppressed when the photoelectrode is composed of a single semiconductor with different band gaps. CdS shows good photovoltaic activity and its band gap can be tuned using particle size effects [8]. Thus the use of single semiconductor (CdS) with different band gaps ensures efficient utilization of the incident spectrum and low internal losses which improves the performance of multiple band gaps CdS photoelectrochemical (PEC) cell.

The aim of this work is to prepare thin film photoelectrodes with different nanostructures of CdS (by chemical bath deposition -CBD) having different effective band gaps, stacked on top of one another, under different conditions in the bath followed by wet chemical etching. Optical, structural and morphological properties are studied for the different films. Their characteristics as multiple band gap photoelectrode PEC cells are analyzed.

2. EXPERIMENTAL DETAILS

2.1 Preparation of multiple band gaps photoelectrode of CdS

For the synthesis of multiple band gaps photoelectrode of CdS, three layers of CdS nanostructures having different band gaps were prepared by chemical bath deposition technique followed by chemical etching. In brief, the three different layers of CdS were prepared onto plastic, glass and titanium (Ti) substrates as follows:

(a) First layer

CdS film was chemically deposited onto the substrate from an alkaline bath (pH~ 12) containing 1M each of $CdSO_4$ and thiourea, at a temperature of 80 °C and a deposition time of 35 min. Ammonia solution was used as complexing agent. After deposition the CdS films were washed with a fine water jet to remove surface adsorbed particles. The freshly prepared films were etched in dilute hydrochloric acid at room temperature (300K). Finally the etched CdS films were heat treated at 100 ^{0}C for an hour.

(b) Second layer

The second layer of CdS was prepared onto first layer of CdS [9]. The second layer of CdS was prepared from similar aqueous alkaline bath (pH~12) with deposition time 20 min, at a temperature of 60 ^{0}C. Chemical etching of the as-grown sample has performed in dilute hydrochloric acid at 60 ^{0}C. After etching such a film shows flower-like structure of CdS. This double layer of CdS was heat treated at 100 ^{0}C for an hour. The double layer of CdS now behaves as substrate for third layer.

(c) Third layer

The third layer of CdS nanowires was prepared onto double layer of CdS [10]. In brief, nanocrystalline sample was prepared from similar aqueous alkaline bath (pH~12) with deposition time 5 hours, at room temperature (300K). The substrate coated with clusters of CdS nanoparticles was immersed immediately in dilute hydrochloric acid at low temperature (~ 283K) in order to obtain nanowires. Finally, the whole sample (triple layer) was heat treated at 100 ^{0}C for an hour before characterization and photoelectrochemical (PEC) measurement.

2.2. Characterization techniques

The surface morphology of the films was characterized by scanning electron microscopy (SEM; JEOL-JSM 5600). The optical transmission data in the range 400–700 nm were obtained with Perkin Elimer, Lambda-35 spectrometer. The PEC cell was fabricated using a two electrode configuration, comprising n-CdS thin film as photoelectrode and graphite rod as a counter electrode. The redox electrolyte was an aqueous solution of 1M NaOH + 1M Na_2S + 1M S. The distance between photoelectrode and counter electrode was fixed to 10 mm with plastic spacer. The photoelectrochemical properties of CdS photoelectrodes were measured under 100 mW/cm^2 light illumination intensity by a 50 W halogen-tungsten filament lamp.

3. RESULTS AND DISCUSSION

The thickness of CdS films were measured with the help of weight difference (gravimetric) method employing sensitive electronic microbalance. Thickness of the CdS film grown at 80 ^{0}C is calculated to be 1375 nm whereas it is 451 and 435 nm for 60^{0}C and 27^{0}C CdS sample respectively.

3.1 Morphology and formation mechanism of CdS nanostructures

Fig. 1 represents the SEM images of as-grown CdS samples fabricated by CBD at 80 ^{0}C, 60 ^{0}C and 27 ^{0}C respectively. The SEM images show that surface is smooth and well covered with CdS film. Small grains of spherical shaped CdS particles can be seen on the surface of films. After wet chemical etching the morphology of samples is changed drastically into different nanostructures of CdS (Fig. 2). After chemical etching the first layer of as-grown sample (prepared at CBD 80 ^{0}C) show honeycomb structured morphology where as flower-like and nanowire morphology is exhibited by second layer (prepared at 60 ^{0}C) and third layer (prepared at 27 ^{0}C) respectively.

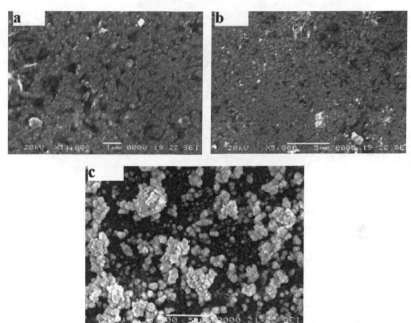

Fig. 1. SEM images of as-grown CBD-CdS samples prepared at (a) 80 ^{0}C (b) 60 ^{0}C and (c) 27 ^{0}C.

Such morphology is possible due to the dissolving tendency of CdS in acidic solution and then diffuses through the solvent and gets deposited onto the preferential surface. We believe that the formation of various nanostructures from the CdS clusters is by dissolution-condensation (recrystallization) process [11].

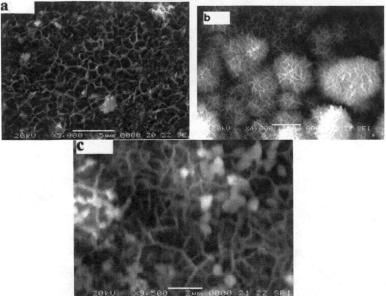

Fig. 2. SEM images of CBD-CdS samples followed by chemical etching in 0.1 M HCl for 5 s prepared at different deposition temperatures and etchant temperatures (a) $80^0C/27^0C$ (b) 60 $^0C/60^0C$ and (c) $27^0C/10^0C$.

Morphology of any semiconductor depends on etching conditions. In the etching process optimized etching parameters should be require for one-dimensional growth of semiconductors. High concentration or high temperature of hydrochloric acid could form highly supersaturated solution of the CdS nanocrystal leading to formation of amorphous or powder of CdS. These dissolve CdS species and effectively terminate the single crystal growth. A low supersaturation is required for anisotropic growth in dissolution-condensation process. It has been reported that the film prepared at high temperature by chemical deposition technique has crystalline nature whereas the film prepared at low temperature has amorphous nature [12]. Therefore in our experiments, the temperature of etchant kept low for the third layer of CdS which was prepared at low temperature (300K) for low supersaturation of CdS species. However some CdS particles did not develop into fibril- shaped structures and kept growing into larger colloids that were also stable (Fig. 2a).

In case of flower-like sample (second layer) high temperature of etchant is the main cause to produce such morphology of semiconductor. In our opinion, the higher temperature of etchant may enhance the chemical kinetics of diffusion of dissolved CdS molecules. Due to this effect not only interconnected short nanorods are formed but also these nanorods grouped with each other to form flower-like nanocrystal of CdS. SEM micrograph of multilayered sample is also depicted in Fig. 3. The nanostructures of CdS well covered the substrate. It can be seen from the figure that there is no uniformity of nanostructures of CdS, i.e. some complicated structures are appeared.

Fig. 3. SEM images of CBD-CdS followed by chemical etching of multilayered samples

3.2 Optical characterization

Optical studies were performed by measuring the transmittance of the films deposited onto plastic and glass substrates in the wavelength range 400-700 nm. Fig. 4 shows the transmission spectra of films of nanowire, flower-like and honeycomb structured CdS. The band gap has been calculated by extrapolating the linear region of the plots $(\alpha h v)^2$ vs. $h v$ on the energy axis, as shown in Fig.5. The band gap values are calculated to be 2.48, 2.40 and 2.22 eV for nanowire, flower-like and honeycomb structured CdS samples respectively.

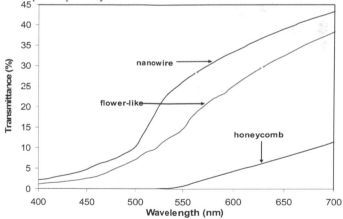

Fig. 4: Optical transmission spectra of three different CdS layers.

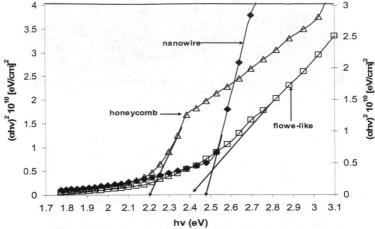

Fig. 5: Plot to determine the direct band gap of the three different CdS layers.

3.3 Photoelectrochemical studies

Photovoltaic output characteristics of nanowires, flower-like, honeycomb structure and multiple band gaps photoelectrode of CdS, under the illumination intensity 100 mW/cm^2, are shown in Fig. 6. The performances of nanowire, flower-like, honeycomb structure and multiple band gaps of CdS PEC cells are summarized in Table I.

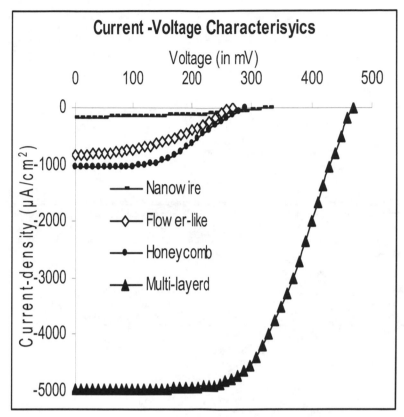

Fig. 6. Photovoltaic output characteristics for n-CdS/1M (NaOH-Na$_2$S-S)/C PEC cells of (a) Nanowire (b) flower-like (c) honeycomb and (d) multiple band gap of all these three layered CdS photoelectrode.

Table I
PEC Cell Parameters

Photo-electrode	Voc (mV)	Jsc (μA/cm^2)	FF (in %)	Effi. (in %)
Nanowire (NW)	330	181	40.0	0.024
Flower-like (FL)	270	827	40.8	0.091
Honey-comb (HC)	290	1034	47.5	0.142
NW/FL/HC triple layer	470	4977	58.4	1.365

The short-circuit current density (J_{sc}), open-circuit voltage (V_{oc}), fill-factor and efficiency of the multiple band gaps CdS PEC cell are larger than that of all the single band gap CdS PEC cells. The short circuit current is low for CdS nanowires structure, increases for flower-like structure and maximum for honeycomb structure. Just opposite is seen for open circuit voltage. By using all the three layers, one over other, as photoelectrode, the short circuit current is 5 times than single honeycomb structured layer and open circuit voltage is 1.5 times than single nanowire structured layer. The efficiency is found to be very low for nanowire structure and maximum for honeycomb structure in case of single layered photoelectrodes. For multi-layered photoelectrode, efficiency is about 10 times as compared to single honeycomb structured photoelectrode. The dense networks of nanowires and flower-like CdS are beneficial in terms of direct conduction paths for efficient electron collection. Nanowire, flower-like and honeycomb morphologies provide high porosity for efficient permeability of electrolytes into the inner structure while maintaining a high surface area for enhanced surface activities. This would then promote prompt carrier separation to obtain good collection efficiency.

4. CONCLUSIONS
    The studies have shown that just by changing the preparation conditions and etching and annealing treatments various types of nanostructures of single material can be grown having different effective band gaps. Use of multi-layers photoelectrode in PEC cells, cell parameters can be enhanced and efficiency can be increased. The multi-layered tandem arrangement ensures much better match to solar spectrum and provides direct conduction path for charge carriers.

REFERENCES
[1] B. Tian, X. Zheng, T.J. Kempa, Y. Fang, N. Yu, G. Yu, J. Huang, C.M. Lieber, Nature, 449 (2007) 885.
[2] O.V. Galán, A.A. Carbajal, R.M. Pérez, G. Santana, J.S. Hernández, G.C. Puente, A. M. Acevedo and M.T. Velázquez, Sol. Ene. Mate. Sol. Cells, 90 (2006) 2221.
[3] C. Heske, D. Eich, R. Fink, E. Umbach, T. van Buuren, C. Bostedt, L.J. Terminello, S. Kakar, M.M. Grush, T.A. Callcott, F.J. Himpsel, D.L. Ederer, R.C.C. Perera, W. Riedl and F. Karg, Appl. Phys. Lett. 74 (1999) 1451.
[4] S.S. Kale, U.S. Jadhav and C.D. Lokhande, Bull. Electrochem. 12 (1996) 540.

[5] H. Jia, Y. Hu, Y. Tang, L. Zhang, Electrochem. Commu. 8 (2006) 1381.

[6] A.B.M.O. Islam, N.B. Chaure, J. Wellings, G. Tolan, I.M. Dharmadasa, Mater. Charact., 60 (2009) 160.
[7] F. Dimroth, Phys. stat. sol. (c) 3, No. 3 (2006) 373.
[8] R. Seigel, Scientific American, 275 (1996) 74.
[9] J.K. Dongre, M. Ramrakhiani, J. Alloy and comp., 487 (2009) 653.
[10] J.K. Dongre, V. Nogriya, M. Ramrakhiani, App. Sur. Sci., 255 (2009) 6115.
[11] G. Cao, Nanostructure and Nanomaterials, first ed., Imperial College Press, London, 2004.
[12] H. Moualkia, S. Hariech, M.S. Aida, Thin Solid Films, 518 (2009) 1259.

EVALUATION OF NANOPARTICLES AS CONTRAST AGENT FOR PHOTOACOUSTIC IMAGING IN LIVING CELLS

Yvonne Kohl and Hagen Thielecke
Fraunhofer Institute for Biomedical Engineering, Division of Biohybrid Systems
Ensheimer Straße 48, 66386 St. Ingbert, Germany

Wolfgang Bost and Robert Lemor
Fraunhofer Institute for Biomedical Engineering, Division of Ultrasound
Ensheimer Straße 48, 66386 St. Ingbert, Germany

Frank Stracke
Fraunhofer Institute for Biomedical Engineering, Division of Biomedical Optics
Ensheimer Straße 48, 66386 St. Ingbert, Germany

Christian Kaiser, Michael Schroeter, and Karl Kratz
Centre for Biomaterial Development, GKSS Research Centre Geesthacht
Kantstraße 55, 14513 Teltow-Seehof, Germany

Andreas Henkel and Carsten Sönnichsen
Johannes Gutenberg University Mainz, Institute for Physical Chemistry
Welderweg 11, 55099 Mainz, Germany

ABSTRACT

Photoacoustic imaging is a hybrid imaging modality combining beneficial features from optical and ultrasound techniques. In photoacoustic imaging, the thermoelastic effect is used for laser-induced generation of ultrasound waves. Absorption of light in tissue leads to small increase in temperature which results in volume expansion and generation of pressure waves. For contrast enhanced imaging, different types of contrast agents can be used. Depending on the geometrical shape plasmon resonance material shows high absorbing properties in the near infrared (NIR). The absorption maximum of existing nanoparticulate contrast agents is located in the range between 700 nm and 900 nm. Due to the limited number of technically available lasers emitting in this spectral range as well as the controversial discussed questions about cytotoxicity, the widespread clinical use is limited. In order to allow a future use of photoacoustic imaging in the clinical routine, laser systems combining the availability of a near-infrared (NIR) wavelength with cost-efficiency and easy-handling such as Nd:YAG lasers (1064nm laser beam) combined with a bioconjugated contrast agent have to be used for signal generation.
The aim of this study was to synthesize gold nanoparticles and infrared dye (IR5)-loaded PLGA (poly(D,L-lactide-co-glycolide)) particles with an absorbance maximum in the range of 1100 nm, to be compatible with the laser system, and to characterize their cytotoxicity with regard to the application as contrast agents for photoacoustic imaging. Using human liver carcinoma cells (HepG2) *in vitro* experiments were performed after four different exposure times to guarantee the biocompatibility of the synthesized nanoparticles. The results indicate that both nanoparticulate systems induce no cytotoxic effects. Using the developed highly sensitive high-resolution photoacoustic microscope platform based on the SASAM acoustic microscopy system concentration-dependent *in vitro* detection of nanoparticles were performed resulting in clear images.

INTRODUCTION

Due to the intrinsic properties of nanostructured materials, like biological barrier transfer, nanoparticles are explored to be used in the medical field.[1,2] At the current state of art nanoparticles

91

are planned to be used as drug delivery -, controlled release - and cell tracking systems, for example in stem cell research. Beside these applications nanoparticulate material, especially metal nanostructures work as contrast agent for medical diagnosis.[3,4] Due to their strong signal per molecular recognition site nanoparticles are of great interest, especially as contrast agents for diagnostic molecular imaging. Particulate contrast agents are currently approved for ultrasound imaging and magnetic resonance imaging.

Beside ultrasound imaging, photoacoustic imaging represents a more sensitive method to be used for diagnosis. Photoacoustic imaging is an emerging non invasive radiation free imaging technique combining beneficial features from optical and ultrasound techniques.[5,6] In this modality, ultrasound signals are generated by absorption of pulsed laser radiation according to their elastic effect. This modality allows acoustical imaging of biological structures with optical contrasts and supports the use of nanoscaled contrast agents with strong optical absorption in the context of molecular imaging. Several biocompatible particle systems with encapsulated Indocyanine green (ICG) have been developed for bio imaging and therapy.[7] Their maximum optical absorption is located at 750 nm wavelength. In this spectral range only cost-intensive and technically complex laser systems can be used for photoacoustic signal generation. In contrast to the cost-intensive OPO laser systems, Nd:YAG lasers emitting 1064 nm light are widely used in scientific, medical and industrial routine and combine cost-efficiency with technically relevant features such as stability and easy handling. Accordingly, the optical absorption properties of photoacoustic contrast agents have to be tuned so that their spectral maximum corresponds to the wavelength of the laser used for signal generation. The spectral absorption of gold nanoparticles depends highly on their size and aspect ratio. Due to the high resonant surface plasmon oscillation gold nanoparticles are used for several applications. In the medical field gold nanoparticles attracted much attention as photothermal agent in hypothermia, as drug-delivery agents and biosensors.[8] Due to its biocompatibility and biodegradability poly(D,L-lactide-co-glycolide) (PLGA) is one of the most utilized materials in the biomedical field and used as scaffolds in bio engineering, implants and as particulate drug delivery systems.[9] Nanoincapsulation of near infrared dyes in a Food and Drug Administration (FDA) accredited material like PLGA is a new approach in particle preparation. The perchlorate IR5 exhibits such absorption maximum, however no information about cytotoxicity and metabolism is yet known. The approach of this study involves the development of two different groups of nanoparticles with an absorption maximum at the relevant laser wavelength of 1064 nm which is widely accessible in common laser systems. With regard to future in vivo applications the cytotoxic effects of the particle systems were determined.

EXPERIMENTAL SECTION

Preparation of PEG-functionalized gold nanorods

Gold nanorods with an absorption maximum at 1064nm were prepared as described in literature.[10,11] The particles are grown by a seed-mediated synthesis at room temperature. Typically, a growth solution containing 5×10^{-4}M HAuCl$_4$, 0.01M cetyltripropyl-ammonimum bromide (CTPAB), 4×10^{-5}M AgNO$_3$ and 7×10^{-4}M ascorbic acids prepared. The particle growth starts upon the addition of seed particles - small preformed gold particles (diameter 2-4nm) - that are produced separately by reduction of HAuCl$_4$ with NaBH$_4$ in a 0.1M CTPAB solution. CTPAB, a cationic detergent, is used as stabilizing agent during preparation of gold nanorods. The nanorods were modified with polyethylene glycol (PEG) to optimize the biocompatibility.

Preparation of IR5-loaded PLGA nanoparticles

The particles are prepared by spray drying. A solution of Poly (D,L-lactide-co-glycolide) PLGA (Res0281, 50:50) and IR5 (8-[[3-[(6,7-dihydro-2,4-diphenyl-5H-1-benzopyran-8-yl)methylene]-1-cyclohexen-1-yl]methylene]-5,6,7,8-tetrahydro-2,4-diphenyl-1-benzopyrylium perchlorate) suspended in methylene chloride were mixed in a ratio 1:100 and spray dried under inert conditions using the mini spray dryer B-290 (Buchi, Switzerland). The nitrogen spray flow and aspirator rate were kept constant at eight bar. The particles were dissolved in 1% polyvinyl alcohol (PVA) solution. By centrifugation nanoparticle fractions of different size ranges were obtained.

Nanoparticle Characterization

The morphology of the nanostructures was investigated by scanning electron microscopy (SEM) (SUPRA™ 40 VP). Samples were mounted on an aluminium stub, coated with Pt/Pd and analyzed using a electron voltage of 3 kV. The morphology of the gold nanorods was determined determined by transmission electron Microscopy (TEM) using a FEI Tecnai F30 electron microscope operating at 300kV. UV-visible absorption spectra were taken on a two-beam UV/VIS spectrometer (Lambda 950, Perkin Elmer, USA). Before absorption measurements the IR5-loaded PLGA nanoparticles, as well as the pure dye IR5, were diluted in chloroform. The UV-visible absorption spectra, recorded at room temperature, ranges from 300 nm to 1400 nm. Zeta potential measurements were performed using a Malvern Instruments Zetasizer Nano (Malvern Instruments Ltd.), operating with a He–Ne laser at 632 nm. Measurements were taken in zeta cells (DTS 1060C).

Cytotoxicity Experiments

Cytotoxicity studies were performed, using HepG2 cells (ATCC, LGC Promochem, Germany), derived from a human hepatocarcinoma. The cell line is cultured in RPMI 1640 without L-glutamine, supplemented with penicillin/streptomycin, sodium pyruvate, glucose and 10% foetal calf serum (FCS). To monitor the cytotoxic effect of the synthesized nanostructures 10^4 cells/well were seeded in a 96well micro titre plate. At day 2, the adherent cell were washed with PBS and exposed to different nanoparticle-concentrations in the range of 0-10µg/ml for one, two, three and five days. To achieve these final concentrations the prepared nanorods were diluted in cell culture medium. Each experiment included a positive control, which was TritonX-100 1%.

The mitochondrial function of the cells exposed to the gold nanorods was analyzed using the WST-1 assay (Roche Diagnostics). This assay is based on the cleavage of stable tetrazolium salt WST-1 by metabolically active cells to an orange formazan dye. WST-1 assay was performed after 24h, 48h, 72h and five days according to manufacturer's instructions, with appropriate controls. After exposure to nanorods the cells were incubated with the WST-1 reagent for four hours. Thereafter the absorbance was quantified at 650 nm using scanning multi-well spectrophotometer reader (Tecan Deutschland). The measured absorbance directly correlates to the number of viable cells.

The detection of the proliferation rate of the cells exposed to nanorods was performed using the BrdU assay kit (Roche Diagnostics). This colorimetric immunoassay is based on the measurement of BrdU (5-bromo-2'-deoxyuridine) incorporation during DNA synthesis. The reaction product is quantified by measuring the absorbance using a scanning multi-well spectrophotometer.

To detect the membrane integrity of the cells exposed to the gold nanorods, lactate dehydrogenase (LDH) release is monitored. During the LDH assay (Roche Diagnostics), LDH released from damaged cells oxidizes lactate to pyruvate, which promotes conversion of tetrazolium salt to formazan, a water-soluble molecule with absorbance at 490 nm. 24h, 48h, 72h and five days after nanorod exposition the supernatant of the cells was transferred in a new 96well micro titre plate and mixed with the corresponding volume of LDH reagent. The formazan dye was quantified using scanning multi-well spectrophotometer reader. The amount of LDH released is proportional to the number of necrotic cells.

All experiments were replicated three independent times and the data are presented as mean SD (Standard error of mean). For the in vitro studies, each stock solution was diluted serially to yield the different concentrations (0µg/ml-10µg/ml). Each experimental value was compared to the corresponding control value for each time point. Statistical significance versus control (cell culture medium) was established as $p < 0.005$. Statistical tests were performed by one-way ANOVA.

Microscopic Photoacoustic Imaging

Suitability of gold and IR5-loaded nanoparticles for photoacoustic imaging was investigated using a custom designed photoacoustic microscope with spatial resolution in the µm range.

The photoacoustic imaging system (SASAM OPTO) consists of the acoustic microscopy platform (SASAM 1000, kibero GmbH, Germany) (Figure 1A).[12-14]
The platform is developed on an Olympus IX81 optical microscope with a rotating column that has an optical condenser for transmission optical microscopy and an acoustic module (Figure 1B) for the acoustic microscopy. The adapted optoacoustic module consists of a Q-switched Nd:YAG solid-state-laser (Teem Photonics, France) generating sub-nanosecond pulses at kilohertz repetition rates. The solid state laser is coupled to the photoacoustic instrument via single mode fiber (core diameter 5μm) and is focused on the sample using the microscope optics. Each pulse used in this experiment had a pulse energy of 350 nJ and a duration of 700 ps. The confocal arrangement allowed high signal to noise ratio photoacoustic signals (>30 dB) to be detected at approximately 400 MHz. In imaging mode, the full width at half maximum value (FWHM) was measured to be 3.6 μm for the 400 MHz transducer. The photoacoustic microscope scans point-by-point along the sample surface and converts the received time resolved signals into a two dimensional image by using the maximum amplitude projection (MAP) or into a three dimensional image by using the acoustic wave time of flight. With this newly developed photoacoustic microscope it is possible to evaluate the photoacoustic suitability of different kind of absorbing particles.

Different concentrations of the synthesized nanoparticles were embedded in agarose to perform concentration-dependent measurements using the developed photoacoustic microscope. The photoacoustic signal of gold nanoparticles in the range of 0.38 mg/ml and 1.52 mg/ml, and IR5-loaded PLGA nanoparticles in the range of 3.7 mg/ml and 14.8 mg/ml were measured. Afterwards the recorded signal amplitudes of each individual samples were compared.

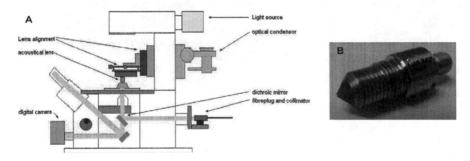

Figure 1. Schematic setup of the photoacoustic microscope based on an inverted microscope. Video or/and visual inspection of the sample is possible. To reflect the excitation light to the objective a short pass dichroic mirror with a cut-off wavelength of 700 nm is integrated. Additionally to the photoacoustic imaging mode, all common optical imaging modalities and pure acoustic microscopy are implemented.

RESULTS AND DISCUSSION
Nanoparticle Characterization
Polyethylenglycol (PEG)-modified gold nanorods were prepared by mixing CTPAB-stabilized gold nanorods with PEG, which reduces their rate for clearance. PEG-modified gold nanorods with an aspect ratio of 5:1 resulting in an absorption maximum in the NIR at 1064 nm were synthesized. Analyzing the TEM images of the PEG-modified gold nanorods (Figure 2A) the size distribution of the nanorods with an absorption maximum at 1064nm (Figure 2B) revealed a mean size between of 39.9 ± 4.4 nm in length and 11.2 ± 2.9 nm in width. The zeta potential of the gold nanorods, without PEG-modification and after PEG-functionalisation, were determined. Gold nanorods stabilized in CTPAB have a cationic surface. Their zeta potential conducts +31.0 mV. This was due to absorbed

CTPAB that has amine as hydrophilic head. PEG-modified gold nanorods showed also a cationic surface (+18.3 mV).

By spray drying IR5-loaded PLGA particles were synthesized (Figure 2B). IR5, with unknown cytotoxic potential, was encapsulated in the PLGA particles by spray drying. The size distribution of the IR5-loaded PLGA particles revealed a mean diameter of 324 nm. The absorption spectra of the pure IR5 dye and the dye-loaded PLGA particles prepared in chloroform was measured (Figure 3). IR5-loaded PLGA particles show a specific absorbance in the NIR at 1100nm, which is comparable to that of the pure dye. The zeta potential of the dye-free PLGA particles has a value of -13.25 ± 0.35 mV and decreases by IR5-loaded particle preparation (-24.05 ± 0.31 mV). This fact could be caused by the IR dye molecules which are located also on the surface of the particles.

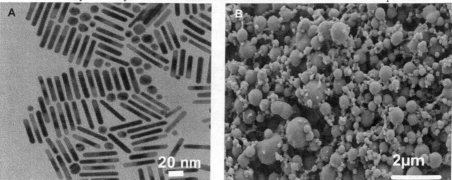

Figure 2. Characterization of the synthesized gold nanoparticles. Transmission electron microscopic images of gold nanorods with a mean size of 40 nm in length and 11 nm in width (A). Scanning electronic microscopic images of IR5-loaded PLGA particles with a diameter in the range of 100 – 600 nm (B).

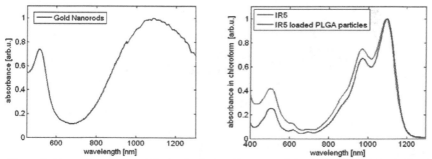

Figure 3. Absorption spectra of the synthesized gold nanoparticles (left), the IR5-loaded PLGA particles and the pure dye IR5 (right). Both nanoparticle systems possess an absorption maximum in the range of 1100nm.

Cytotoxicty Experiments

The cytotoxicity of PEG-modified gold nanorods and IR5-loaded PLGA nanoparticles in human hepatocellular cells (HepG2) was evaluated. HepG2 cells were derived from the human liver, one of the major organs of metabolism and biotransformation and assumed as location of nanoparticle clearance and possible accumulation. Due to these facts, to study the biotransformation and to simulate the metabolism of the nanoparticles in the human body, HepG2 cells were used as *in vitro* cell culture test system. The cytotoxic potential of the synthesized nanoparticles in a concentration range of 0.1 to 10 μg/ml were investigated. The metabolic activity was determined using the WST-1 assay. Following four different exposure times (24h, 48h, 72h and five days), quantitative analyses of cell viability was carried out. The gold nanorods with PEG surface modifiers did not appear to be cytotoxic in HepG2 cells at concentrations up to 10μg/ml. Even after five days of chronically exposure no decrease in metabolic activity was induced (Figure 4A). Also the IR5-loaded nanoparticles are biocompatible up to 10μg/ml using the WST-1 assay (Figure 5A). Via BrdU assay the PEG-modified gold nanorods induced no significant decrease in proliferation rate of the HepG2 cells after 24 h, 48 h and 72 h at the concentration of 10μg/ml, even after 5 days of exposure the proliferation rate remained constant at the level of the control (100%) (Figure 4B). Also the IR5-loaded nanoparticles are biocompatible up to 10μg/ml using the BrdU assay (Figure 5B). Dye-free PLGA particles also appear to be non-toxic in HepG2 cells at concentrations up to 10μg/mL, even after 5 days of chronically exposure (data not shown). Membrane integrity after exposure to the prepared nanoparticles was investigated by the LDH assay. These results give evidence on cell-nanoparticle interaction. Compared to the nanorod-free control no LDH (lactate dehydrogenase) leakage was induced by any of the PEG-modified gold nanorods (Figure 4C) and IR5-loaded PLGA nanoparticles (Figure 5C). The exposed nanoparticles induce a LDH leakage of 20%, which correlates to the noise level of the system. The results of the cytotoxicity studies indicate biocompatibility of the nanoscaled materials prepared within this study.

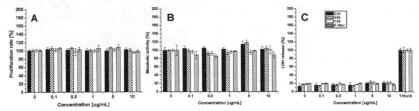

Figure 4. Effect of PEG-functionalized gold nanorods with an absorption maximum at 1064 nm on the viability of HepG2 cells. Cells were treated with different concentrations (0-10 μg/ml) of PEG-modified gold nanorods with an absorbance maximum at 1064 nm. After 24h, 48h, 72h and five days of exposure the effect on cell proliferation (A), metabolic activity (B) and membrane integrity (C) was quantified. Data are expressed as mean ± SD (n = 3). Control cells without gold nanorods treatment are 100 %.

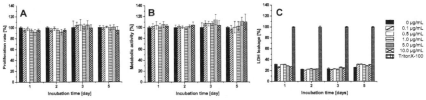

Figure 5. Effect of IR5-loaded PLGA nanoparticles with an absorption maximum at 1100 nm on the viability of HepG2 cells. Cells were treated with different concentrations (0-10 μg/ml) of PEG-modified gold nanorods with an absorbance maximum at 1064 nm. After 24h, 48h, 72h and five days of exposure the effect on cell proliferation (A), metabolic activity (B) and membrane integrity (C) was quantified. Data are expressed as mean ± SD (n = 3). Control cells without gold nanorods treatment are 100 %.

Photoacoustic imaging

This ultra sensitive detection platform allows high resolution imaging of infrared absorbing structures and can especially be used for nanoparticle detection. After embedding different concentrations of the new developed nanoparticles in an agarose matrix photoacoustic images were taken by using the maximum amplitude protection (Figure 6). The signal amplitudes of the two nanoparticle concentrations were displayed (Figure 7). In both cases the detected signal amplitude correlates to the particle concentration. An increasing amount of nanoparticles results in an increase of the signal–to–noise ratio (Figure 7). With the background of the biocompatibility experiments the prepared PEG-functionalized gold nanorods and IR5-loaded PLGA particles seem suitable for photoacoustic diagnostic purpose up to 10 µg/ml. Comparing the signal generation of the highest concentrations of both synthesized nanoparticle systems demonstrates the high photoacoustic signal and indicates the bases for the next step on the way to a new class of contrast agents for photoacoustic imaging.

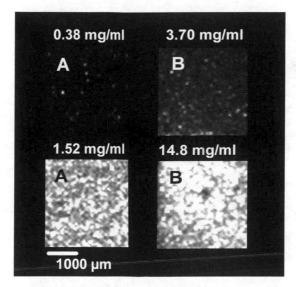

Figure 6. Concentration-dependent photoacoustic imaging of PEG-functionalized gold nanorods (A) and IR5- PLGA nanoparticles (B) embedded in agarose.

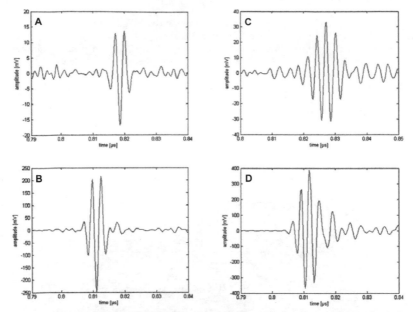

Figure 7. Signal amplitudes of PEG-functionalized gold nanorods (A) and IR5-loaded PLGA nanoparticles (B) embedded in agarose. The signal amplitudes were calculated based on the photoacoustic images.

CONCLUSION

Using nanoparticulate contrast agents with an absorption maximum located in the range of 750 nm limits their widespread clinical use, because in that spectral range only cost-intensive and technically complex OPO laser systems are available for optoacoustic signal generation. Nd:YAG lasers emitting 1064 nm light. Due to the fact that the synthesized PEG-functionalized gold nanorods and the NIR dye IR5 show an absorbance maximum in the range of 1100 nm, the dye-loaded particles represent a particle system which could be used as photoacoustic contrast agent, in combination with a cost-efficient photoacoustic hardware platform.

The current results of the performed *in vitro* experiments indicate no cytotoxic potential of both synthesized particle systems. The results of the performed experiments characterise the PEG-functionalized gold nanorods and the IR5-loaded PLGA particles as a biocompatible nanoparticulate systems for the planned application as contrast agent for photoacoustic imaging. The high optical absorption of the particle systems results in excellent photoacoustical signals comparable to commercially available contrast agents. We have demonstrated for the first time that IR5-loaded PLGA particles present a new class of biocompatible contrast agents for photoacoustic imaging, with absorption in a wavelength regime that allows for higher tissue penetration depth than particle systems proposed by others.

ACKNOWLEDGEMENT

This work was funded by the German Federal Ministry of Education and Research BMBF in the context of the Research Project POLYSOUND (contract number 0312029).

REFERENCES

[1]N. Lewinski, V. Colvin, R. Drezek, Cytotoxicity of nanoparticles., *Small*, **4**, (1), 26-49 (2008).

[2]S.K. Murthy, Nanoparticles in modern medicine: state of the art and future challenges., *Int. J. Nanomedicine*, **2**, (2), 129-41 (2007).

[3] Y.T. Lim, Y.W: Noh, J.H. Han, Q.Y. Cai, K.H. Yoon, B. H. Chung, Biocompatible polymer-nanoparticle-based bimodal imaging contrast agents for the labeling and tracking of dendritic cells., *Small*, **4**, (10), 1640-5 (2008).

[4]J. Kreuter, S. Gelperina, Use of nanoparticles for cerebral cancer., *Tumori*, **94**, (2), 271-7 (2008).

[5]M. Xu, L. V. Wang, Photoacoustic imaging and sensing in biomedicine, *Review of Scientific Instruments*, **77**, 041101 (2006).

[6]C.H. Li, L.V. Wang Photoacoustic thomatography and sensing in biomedicine., *Physics in Medicine and Biology*, **54**, 59-97 (2009).

[7]V. Saxena, M. Sadoqi, J. Shao, Indocyanine green-loaded biodegradable nanoparticles: preparation, physicochemical characterization and in vitro release., *Int J Pharm,* **278**, (2), 293-301 (2004).

[8]G.R. Souza, D.R. Christianson, F.I. Staquicini, M.G. Ozawa, E.Y. Snyder, R.L. Sidman, J.H. Miller, W. Arap, R. Pasqualini, Networks of gold nanoparticles and bacteriophage as biological sensors and cell-targeting agent, *Proc. Natl. Acad. Sci U S A.*, **103** (5), 1215-1220 (2006).

[9]F.Y. Cheng, S.P.H. Wang, C.H. Su, T.L. Tsai, P.C. Wu, D.B. Shieh, Stabilizer-free poly(lactide-co-glycolide) nanoparticles for multimodal biomedical probes. *Biomaterials*, **29** (13), 2104-2112 (2008).

[10]X. Kou, S. Zhang, C.K. Tsung, Z. Yang, M. Yeung, G. Stucky, L. Sun, J. Wang, C. Yan, *Chem. Eur. J.*, **13**, 2929-2926 (2007).

[11]X. Kou, S. Zhang, C.K. Tsung, M.H. Yeung, Q. Shi, G.D. Stucky, L. Sun, J. Wang, C. Yan, Growth of gold nanorods and bipyramids using CTEAB surfactant, *J. Phys. Chem. B.*, **110** (2006).

[12]W. Bost, R.M. Lemor, Photoacoustic microscopy of high resolution imaging. *The Journal of the Acoustical Society of America*, **123** (5), 3370 (2008).

[13]W. Bost, F. Stracke, E.C. Weiss, S. Narasimhan, M.C. Kolios, R.M. Lemor, High frequency optoacoustic microscopy. *Conf. Proc. IEEE Eng. Med. Biol. Soc.*, **1**, 5883-5886 (2009).

[14]E.C. Weiss, P. Anastasidis, R. M. Lemor, P.V. Zinin, *IEEE Vol. 54*, **11**, 2257-2271 (2007).

THE USE OF CaCO$_3$ AND Ca$_3$(PO$_4$)$_2$ AS SUPPORTS FOR Fe-Co CATALYSTS
FOR CARBON NANOTUBE SYNTHESIS: A COMPARATIVE STUDY

Sabelo D. Mhlanga[a,b*], Suprakas Sinha Ray[a] and Neil J. Coville[b]

[a]DST/CSIR Nanotech. Innovation Centre, National Centre for Nano-Structured Materials, P.O. Box 395, Pretoria, 0001, South Africa.
[b]DST/NRF Centre of Excellence in Strong Materials and the Molecular Sciences
Institute, School of Chemistry, University of the Witwatersrand, WITS 2050, South Africa
*Corresponding author. Email: Sabelo.Mhlanga@wits.ac.za, Tel.: +27 11 717 6705, Fax: +27 11 717 6749

ABSTRACT
       The use of a carbon nanotube (CNT) supported catalyst system enhanced with 'docking stations' along the exterior has been shown to limit the surface mobility of ultra small iron catalyst particles on the CNT surfaces during Fisher–Tropsch synthesis. Generally, the mobility of surface-bound metallic nanoparticles on catalyst supports results in sintering, leading to a subsequent decrease in effectiveness of the catalytic behaviour of the metal nanoparticles over time. A Fe-Co bimetallic mixture (50:50 w/w) was impregnated (5 mass % loading) onto either Ca$_3$(PO$_4$)$_2$ to give < 1% multiwalled CNTs (MWCNTs) or onto CaCO$_3$ to give high yields of MWCNTs with smoother surfaces. Mixtures of Ca$_3$(PO$_4$)$_2$-CaCO$_3$ (0/100 to 100/0) yielded tubes with very rough surfaces with larger diameters and the CNT yield increased as the amount of CaCO$_3$ in the support mixture was increased. The results suggest that metal-support interactions supersede surface area. We show that the presence of CO$_2$ generated from the decomposition of CaCO$_3$ assists in the formation of CNTs and avoids thickening of the CNTs. This was confirmed by performing studies using other supports such as CaO, SiO$_2$ and Al$_2$O$_3$. These CNTs should be suitable for use as catalysts supports and other industrial applications including polymer composites and ceramics. The support mixture is thus ideal for the large scale production of the CNTs since the supports are cheap and environmentally friendly.

INTRODUCTION
       Carbon nanotubes (CNTs) are classified as an allotrope of carbon. Other allotropes include diamond, graphite, fullerene and amorphous carbon. The CNTs are one of the strongest and stiffest materials known, in terms of their tensile strength and elastic modulus respectively, which is the consequence of the covalent sp^2 bonds formed between the individual carbon atoms [1-5]. The extraordinary properties of CNTs (e.g. electrical conductivity, thermal stability and mechanical properties) have made these materials potentially useful in many applications in nanotechnology such as electronics, optics, catalysis, biomedicine, fuel cells etc [2,3,6-13]. Since a study by Iijima [14], that identified CNTs, these materials have attracted much attention from researchers within the scientific community and methods to make CNTs have been extensively investigated.

       The synthesis of carbon nanotubes (CNTs) can be done using arc discharge [15,16], by laser ablation [17], by thermolysis in a closed environment at high pressure [18-27], or by chemical vapour deposition (CVD) procedures [28]. The latter method is generally classified into two categories namely, the floating catalyst process, which is a gas phase process, and the substrate CVD process. The substrate CVD process involves the use of a thermally stable material to support metal particles that are responsible for the molecular decomposition of carbon to form the CNTs [13].
The CVD process is generally the most widely used method because it is a low cost process that is industrially scalable, versatile, and allows for the controlled growth of high purity CNTs.

CaCO₃ and Ca₃(PO₄)₂ as Supports for Fe-Co Catalysts for Carbon Nanotube Synthesis

High surface area materials such as SiO_2, Al_2O_3, TiO_2 and zeolites provide excellent supports for catalytic CNT production. However, the nature of these materials does not allow for their easy removal and thus purification of CNTs produced on these supports remains a challenge and often requires several acidic/oxidation steps that result in damage to the CNT graphitic structures. Environmental aspects should also be taken into consideration because the solvents used during the purification process must be disposed. To circumvent these environmental problems, the use of readily removable supports has been studied both by us and others [29-41]. These supports include MgO, $MgCO_3$, CaO, $Ca(OH)_2$, $CaCO_3$ etc.

Calcium carbonate ($CaCO_3$) has proven to be a support of choice for mono- and bimetallic transition metal particles (typically Fe, Ni and Co) because it satisfies both the catalyst activity and environmental aspects in the synthesis of CNTs. Previous studies have revealed the advantages of using this material as a support for CNT synthesis [29-37]. The disadvantage, however, is that the $CaCO_3$ loses CO_2 to give low surface area CaO at T > 800 °C. The presence of this released CO_2 has however been proposed to favour CNT growth [33].

In this paper, we report on the use of a novel material, calcium pyrophosphate, $Ca_3(PO_4)_2$, as well as $CaCO_3$/$Ca_3(PO_4)_2$ mixtures as supports for CNT synthesis. $Ca_3(PO_4)_2$ has a high melting point (1670 °C), a high surface area of 42 m^2/g (compared to $CaCO_3$, 10 m^2/g), is insoluble in water and alcohols, soluble in mild acids and is not expensive [42]. These properties would render the material highly suitable for the CVD synthesis of CNTs.

The rough CNTs look like carbon nanofibres, which can occur par pyrolytic thickening of MWCNTs [43-45]. Our previous studies using $CaCO_3$ as a support for CNT synthesis showed that the presence of defects on the surface of CVD CNTs can act as "nano docking stations" for metal catalyst nanoparticles and thus assist in avoiding the deactivation of catalysts through sintering in catalytic reactions such as the Fischer-Tropsh reaction [46].

EXPERIMENTAL SECTION

The $Ca_3(PO_4)_2$, $Ca_3(PO_4)_2$-$CaCO_3$ (prepared by physically mixing the two materials) and $CaCO_3$ supported Fe-Co catalysts (2.5% Fe, 2.5% Co by mass) were prepared using a wet impregnation process [34]. $Fe(NO_3)_3 \cdot 9H_2O$ and $Co(NO_3)_2 \cdot 6H_2O$ were purchased from Sigma Aldrich and were used as sources of Fe and Co respectively. The metal loading concentrations were confirmed by Inductively Coupled Plasma – Atomic Emission Spectroscopy (ICP-AES) measurements and by quantitative thermogravimetric analysis (TGA) using a Perkin Elmer Pyris 1 TGA. For thermal analysis, the CNTs were decomposed under a continuous flow of air or nitrogen (40 ml/min) at a heating rate of 10 °C/min.

The CNTs were characterized by transmission electron microscopy (TEM) using either a JEOL 100S Electron Microscope or a Philips CM200 equipped with a Gatan Imaging Filter for higher magnifications and energy dispersive X-ray spectroscopy (EDS). BET surface area analysis (Micromeritics TriStar Surface Area and Porosity Analyzer), Raman spectroscopy (Jobin-Yvon T64000 micro-Raman spectrometer), X-ray photoelectron spectroscopy (XPS) (Physical Electronics Quantum 2000) and powder X-ray diffraction (PXRD) (Bruker axs D8 Advance PXRD) techniques were also used in the study.

MWCNTs were grown using a horizontal fixed bed reactor consisting of an electronically controlled furnace and a quartz tube reactor [34]. From our previous studies [34], it was observed that the activity of Fe-Co catalysts was largely influenced by the reaction temperature and the ratios of nitrogen/acetylene gas mixtures. In this study the synthesis of CNTs was performed varying only the temperature (600 – 1000 °C) using a C$_2$H$_2$:N$_2$ flow rate of 1:3. Experiments were also carried out using C$_2$H$_2$ without a carrier gas (N$_2$) in order to determine the activity of the catalyst under this condition. The role of oxygen was investigated by varying the amount of CO$_2$ generated from the decomposition of the CaCO$_3$ support during the CNT synthesis. The effect of synthesis time on the CNT structure and yield was investigated.

Purification of the CNTs was carried out using a single step process that involved the use of mild acid (5, 10, 30% HNO$_3$) to dissolve and remove the support and metal particles from the final product. This was achieved by stirring continuously at room temperature the mixture of the CNTs and acid (e.g. 3 g in 100 ml) for 30 min - 24 h. The product was washed with distilled water until the washings were neutral.

RESULTS AND DISCUSSION
The Fe-Co catalysts supported on Ca$_3$(PO$_4$)$_2$, CaCO$_3$, and Ca$_3$(PO$_4$)$_2$-CaCO$_3$ mixtures were evaluated for CNT production.

BET surface area analysis
Table 1 shows the details of the catalysts used, the amount of raw product obtained and the average diameters of the CNTs produced. The BET surface area analysis of the catalyst-support materials showed that the total surface area of the materials decreased as the amount of CaCO$_3$ added to Ca$_3$(PO$_4$)$_2$ was increased from 5 – 95% by weight. The decrease in the surface area was expected since CaCO$_3$ has a much lower surface area (10.5 m^2/g) than Ca$_3$(PO$_4$)$_2$ (41.6 m^2/g).

Table 1: Chemical composition of the catalysts used (200 mg) for the synthesis of carbon nanotubes (synthesis time = 1 h) at 700 °C.

Reaction	Catalyst composition					Product analysis	
	Ca$_3$(PO$_4$)$_2$ wt%	CaCO$_3$ wt%	Fe^{3+} wt%	Co^{2+} wt%	BET SA (m^2/g)	Yield (g)	Out. diam. (nm)[c]
1	100	0	2.5	2.5	41.6	0.29	-
2	95	5	2.5	2.5	39.6	0.40	50
3[a]	50	50	2.5	2.5	29.2	0.86	50
4	5	95	2.5	2.5	13.8	1.03	50
5[b]	0	100	2.5	2.5	10.5	1.15	25

[a]See Fig. 1b for TEM image
[b]See Fig. 1a for TEM image
[c]± 5 nm for all samples

The Ca$_3$(PO$_4$)$_2$ catalysts were tested for the synthesis of CNTs under the conditions described in the experimental section. As can be observed (Table 1), reactions over pure Ca$_3$(PO$_4$)$_2$ gave very little carbon deposit that consisted of CNTs (< 1%) and some amorphous carbon. Reactions over CaCO$_3$ as expected gave a high yield of CNTs with near 100% selectivity (Table 1) and no amorphous carbon. The decomposition of C$_2$H$_2$ was then performed using CaCO$_3$ physically mixed with varying amounts of Ca$_3$(PO$_4$)$_2$ (5, 50 and 95 wt%) prepared using the same impregnation procedure. With the Ca$_3$(PO$_4$)$_2$-CaCO$_3$ supports, the CNT yield decreased with increase of Ca$_3$(PO$_4$)$_2$ in the mixture. This observation is explained later in this article.

TEM analysis of the CNTs

TEM analysis was performed on the product obtained from the Fe-Co on the different supports. The TEM analysis revealed that the CNTs produced were all multiwalled in nature with average outer diameters in the range 20 - 50 nm (synthesis time = 1 h). The inner diameters of the CNTs were in the 5 – 10 nm range. The Fe-Co/Ca$_3$(PO$_4$)$_2$-CaCO$_3$ catalysts produced CNTs with outer diameters twice the size of the CNTs obtained using 100% CaCO$_3$. The TEM images showed that the CNTs synthesized in the presence of Ca$_3$(PO$_4$)$_2$ possess extremely rough (pitted) surfaces when compared to CNTs produced from CaCO$_3$ (Figure 1) under the same synthesis conditions. Further, the purified CNTs synthesized from Ca$_3$(PO$_4$)$_2$-CaCO$_3$ have high surface areas ~ 132 m^2/g compared to the CNTs synthesized over CaCO$_3$ (SA = 94 m^2/g), due to the porous nature and roughness of the CNTs.

CNTs with a rough outer wall structure were observed by Escobar et al when CNTs were synthesized using Fe on a SiO$_2$ susbtrate [47]. The amorphous noncrystalline surface of the CNTs was attributed to the concentration of acetylene in the synthesis. At high acetylene concentrations carbon nanoparticles grew, covering the surface of CNTs, giving a compact coating [47]. However, it could be seen here that CNTs synthesized over CaCO$_3$ did not have such an amorphous layer carbon.

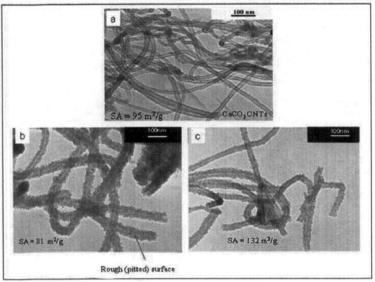

Figure 1. A TEM image of MWCNTs synthesized over CaCO$_3$ (a) and TEM images of MWCNTs synthesized over 50wt%CaCO$_3$/50wt%Ca$_3$(PO$_4$)$_2$ before (b) and after purification (c) with 30% HNO$_3$.

A high magnification TEM (HMTEM) image of the CNTs synthesized over Ca$_3$(PO$_4$)$_2$/CaCO$_3$ (50:50 wt%) with a rough surface is shown in Figure 2. A closer inspection of the outer surface revealed that the surface is covered with an amorphous layer of carbon. The amorphous nature of the CNTs has been observed in several other studies: Luo et al [48] synthesized long amorphous CNTs by a solvothermal treatment, in which ferrocene and sulfur powder were the reactants and benzene served as a carbon source. The CNTs were entirely amorphous. This was also reported in a similar study by Xiong *et al* [49]. Nishino and co-workers [50] also synthesized amorphous CNTs using poly(tetrafluoroethylene) and ferrous chloride. Amorphous CNTs have also been produced in large scale by temperature-controlled arc discharge in a hydrogen atmosphere with Co/Ni alloy powders [50].

The CNTs produced in this study are not entirely amorphous. The random orientation of carbon is limited to the outer part of the CNT walls as shown by the HMTEM image in Figure 2 (region 2). Clearly the inner wall is graphitized and the amount of defects in this region is low, as shown Figure 2, region 1. The graphite sheets are arranged in a perfect orientation contrary to the outer layer shown in region 2. An X-ray diffraction pattern of region 2 revealed the semi-crystalline nature of the nanotubes.

Escobar *et al* [47] proposed that during the synthesis of CNTs, amorphous carbon nanoparticles nucleate on the external wall of the CNTs. The mechanism of this observation has not been explained. However we believe that the deposition of carbon occurs by a mechanism similar to the nucleation of carbon black soot that takes place during the direct pyrolysis of hydrocarbons (e.g. CH_4, C_2H_2, C_2H_4) in the absence of a catalyst at high temperatures [52].

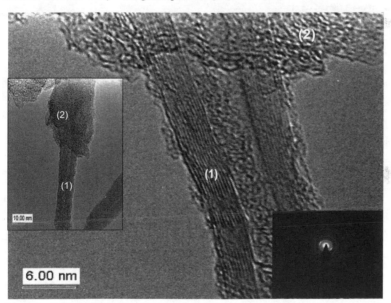

Figure 2. A HMTEM image of CNTs synthesized over CaCO₃-Ca₃(PO₄)₂ (50:50 wt%) after 1 h showing the perfect orientation of graphite sheets of the inner tubes (region 1) and the amorphous part of the CNTs (region 2). An X-ray diffraction pattern of region 2 is shown (inset).

The role of CO_2

It can clearly be seen from the results in Table 1 that $CaCO_3$ is a superior support for CNT synthesis compared to $Ca_3(PO_4)_2$. Small amounts of $CaCO_3$ added to $Ca_3(PO_4)_2$ improved the yield of CNTs. This observation suggested that $CaCO_3$ played important role in the CNT synthesis and the chemistry involved was investigated by considering the effect that may be caused by the decomposition of $CaCO_3$ to CaO resulting to generation of CO_2. Our observation was in agreement with studies by others, which have shown that the presence of a small amount of a species that contains oxygen atoms in addition to the carbon source dramatically improves the yield of the reaction in the CVD synthesis of the CNTs [53-55]. Indeed Magrez et al, has shown that when $CaCO_3$ is used as support, CO_2 is generated in-situ when the $CaCO_3$ decomposes at the synthesis temperature, 700°C. The CO_2 is assumed to act as etching agents to prevent encapsulation of catalyst particles by amorphous carbon [33].

When $Ca_3(PO_4)_2$ is used as a support no oxygen atoms are generated and when equal amounts of $Ca_3(PO_4)_2$ and $CaCO_3$ are used, the amount of oxygen in the form of CO_2 is reduced by 50% and the yield and quality of the CNTs produced are reduced from those obtained when 100% $CaCO_3$ was used as a support. The amount of CO_2 generated was confirmed by performing TGA analysis (Figure 3) of the supports heated in N_2. The TGA profiles indicate that the amount of CO_2 correlates with the amount of $CaCO_3$ in the support mixture.

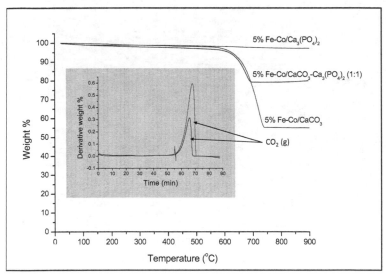

Figure 3. TGA profiles and derivative weight profiles (inset) of the catalyst support mixtures heated in nitrogen.

Indeed, the results shown above suggest that the CO_2 generated from the $CaCO_3$ play an important role in the pyrolysis of C_2H_2 to produce CNTs. To verify this effect on the yield, small amounts of CO_2 were mixed with C_2H_2 over these supports as well as over CaO, SiO_2 and Al_2O_3 to produce CNTs under the same reaction conditions. Small amounts of CO_2 were observed to enhance the CNT formation as shown by the increase in the yield. The amount of amorphous carbon on the surface of the CNTs was found to be suppressed with time thus the CNT yield could be improved without affecting much, the quality of the CNTs. Others have shown that the presence of small amount of CO$_2$ and H$_2$O vapour in the reaction system resulted in an increase in CNT length [55].

Effect of reaction time
It was observed that the outer diameters of the CNTs increase with increase in reaction time and break into shorter tubes after 2 h. (Figure 4c,d) It can thus be said that the increase in CNT synthesis time using the mixture of supports, not only results in good yield and roughening of the CNTs, but also causes fragmentation of the CNTs [56]. The fragmentation of the CNTs can be associated with the deposition of carbon on the already formed CNTs, which results in their fragmentation due to stress. Indeed many of the fragments revealed an amorphous carbon layer deposited on an inner crystalline tube as seen in Figure 4c,d (see circled regions). If the tube has defects, the defects will provide a point for tube rupture. We suggest that it is the defects on the CNTs that cause the CNTs to fragment after carbon deposition on the tubes. These could be associated with the presence of pentagons and heptagons along the tube length [56].

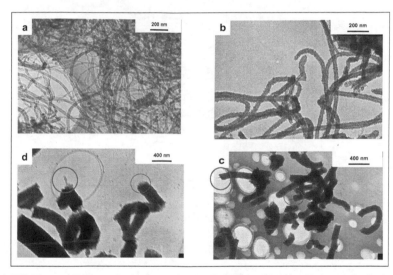

Figure 4. TEM images CNTs synthesized over CaCO$_3$-Ca$_3$(PO$_4$)$_2$ (50:50 wt%) for (a) 5 min (b) 60 min (c) 3 h and (d) 6 h synthesis time.

The rough surface of the CNTs was also measured by Raman analysis (Figure 5). The two peaks occurring at 1348 and 1597 cm^{-1}, corresponding to the D and G-vibration modes of the graphite sheets can be used to indicate the degree of defects/disorder of the graphite sheets. The degree of disorder is given by the ratios of the intensities of the D and G-bands (I_D/I_G). The I_D/I_G ratios of the MWCNTs synthesized after 5 and 60 min were ~ 0.8 and ~ 1.0 respectively suggesting that the disorder of the graphite sheets was very high and increased with exposure to C$_2$H$_2$. Raman analysis also was carried out on the purified (24 h, r.t., 30% HNO$_3$) CNT samples. The I_D/I_G ratios purified

MWCNTs (synthesized in 1 h) were ~ 1 the same as that measured for the unpurified CNTs. Thus the acid treatment process did not affect the graphitic nature of the wall structure of the CNTs.

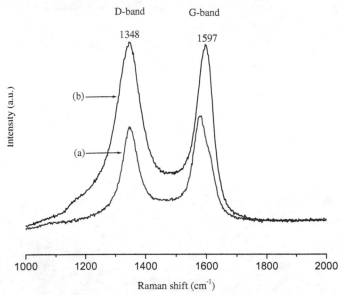

Figure 5. Raman spectra of MWCNTs synthesized over CaCO$_3$-Ca$_3$(PO$_4$)$_2$ (50:50 wt%) for (a) 5 min and (b) 60 min.

Elemental composition and purification

The elemental composition of the as-synthesized CNTs was carried out using XPS analysis. The CNT wall structure was shown to contain C (96.3%) and O (3.3%) (Figure 6). The spectrum of the raw and purified CNTs revealed that the materials are composed mainly of carbon. The elemental composition of the materials was also confirmed by EDS. The elemental analysis revealed that the CNTs are relatively pure and contain undetectable amounts (< 0.5%) of Ca or P implying that large particles of CaO or Ca$_3$(PO$_4$)$_2$ are not embedded in the CNT walls. It is, however, possible that small particles < 3 nm could be deposited in the nanotubes walls [57]. No large crystalline particles of Ca, P or CaO were observed by TEM.

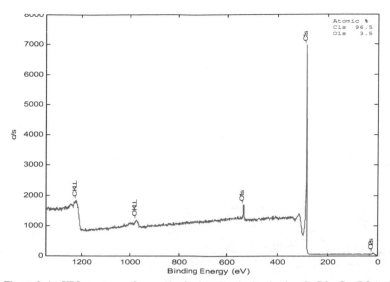

Figure 6. An XPS spectrum of as-synthesized CNTs obtained using CaCO$_3$-Ca$_3$(PO$_4$)$_2$ (50:50 wt%).

The effectiveness of the purification process was also studied using PXRD and TGA (Figure 7). The XRD pattern of the unpurified and purified CNTs showed two peaks at 26° and 45° corresponding to carbon peaks. The carbon peaks are broad; an observation attributed to the low crystallinity of the outer walls of the CNTs [58]. The as-synthesized and purified MWCNTs were also characterized by TGA in air (Figure 7a). The weight loss due to the combustion of carbon with oxygen, which corresponds to the carbon content in the sample, occurred in the temperature range of 550 – 600 °C. The TGA of the as-synthesized and purified CNTs enabled the mass of impurities and CNTs in the carbon deposit to be determined. No amorphous carbon or any other form of shaped carbon nanomaterials were formed in the production process. The TGA profiles are similar to those of MWCNTs obtained using typical CVD processes [31,38]. Very dilute solutions of HNO$_3$ (5, 10, 15%) were also found to be effective in removing the support and metal particles, thus the use of concentrated acids and multi-step purification could be avoided. As anticipated, the higher the concentration (e.g. 15 %) of the acid used, the purification process was seen to be more effective in removing the impurities. The decomposition of the CNTs occurred over a narrow temperature range suggesting increased purity.

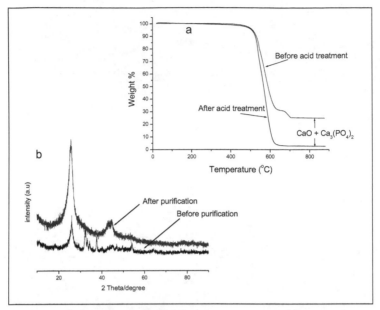

Figure 7. (a) TGA profiles and (b) PXRD pattern of CNTs synthesized over CaCO$_3$-Ca$_3$(PO$_4$)$_2$ (50:50 wt%) before and after purification with 30% HNO$_3$ for 24 h.

CONCLUSION

The work presented herein reports a novel approach for the synthesis of 'rough' MWCNTs in high yields. While these CNTs possess a rough surface, TEM analysis revealed that the CNT inner walls possess a regular array of crystalline graphite sheets (typically 3 - 5 nm in length). Raman analysis revealed that the rough CNTs synthesized in 1 h contained a high degree of defects on the graphite sheets. The study showed that the growth of CNTs over longer synthesis times (t > 2 h) results in thickening and fragmentation of the CNTs due to the deposition of carbon on the formed CNTs.

We demonstrate that the yield and surface roughness of the CNTs can be modified by controlling the amount of CO$_2$ in the reaction system. Hence, by varying the amount of CaCO$_3$ in the support mixture, the amount of CO$_2$ could be varied. We show that when CO$_2$ is supplied into the reaction system, it also enhances the yield and quality of the CNTs. The CO$_2$ also prevents the formation of amorphous carbon. The support materials as well as their mixtures used in this study can readily be removed from the CNTs by dissolving them in very dilute acid (5% HNO$_3$). This purification process does not affect the graphitic structure of the CNTs. The support mixture is thus ideal for the large scale production of CNTs with a rough outer surface. The rough surfaces 'pits' of the CNTs could serve as 'docking stations' for metal particles in catalytic reactions such as the Fischer-Tropsch synthesis, thus reducing deactivation of catalyst through sintering of the metal particles.

ACKNOWLEDGEMENT
We acknowledge financial support from the University of the Witwatersrand, the CSIR (Pretoria), and the Mellon Postgraduate Mentoring Programme (Wits).

REFERENCES
1. M.S. Dresselhaus, G. Dresselhaus and P.C. Eklund, *Science of fullerenes and carbon nanotubes*, Calif: Academic press, San Diego, 1996.
2. M.S. Dresselhaus, G. Dresselhaus and P. Avouris. *Carbon nanotubes: synthesis, structure, properties, and applications*, Springer, Berlin, 2001.
3. P.J.F. Harris. *Carbon nanotubes and related structures: new materials for the twenty-first century*, Cambridge University Press, Cambridge, U.K., New York, 2001.
4. R. Saito, G. Dresselhaus and M.S. Dresselhaus, *Physical properties of carbon nanotubes*, Imperial College Press, London, 1998.
5. S. Reich, C. Thomsen and J. Maultzsch, *Carbon nanotubes: basic concepts and physical properties* Wiley-VCH, Weinheim, Cambridge, 2004.
6. B.M. Endo, T. Hayashi, Y.A. Kim, M. Terrones, M.S. Dresselhaus, *Carbon nanotubes in the twenty-first century one contribution of 12 to a Theme 'Nanotechnology of carbon and related materials'*, Phil. Trans. R. Soc. Lond. A 362 (2004) pp. 2223-2238.
7. P. Ball, *Made to measure: New materials for the 21st Century*, University Press, Princeton, USA, 1997.
8. M. Terrones, A. Jorio, M. Endo, Y.A. Kim, T. Hayashi, H. Terrones, J-C. Charlier, G. Dresselhaus and M.S. Dresselhaus, *New direction in nanotubes science*, Mater. Today 7 (2004) pp. 30-45.
9. M. Paradise and T. Goswami, *Carbon nanotubes – Production and industrial applications*, Mater. Design 28 (2007) pp. 1477-1489.
10. J.K. Ong, N.R. Franklin, C. Zhou, M.G. Chapline, S. Peng, K. Cho and H. Dai, *Nanotube molecular wires as chemical sensors*, Science 287 (2000) pp. 622-625.
11. S.J. Tans, R.M. Verschueren and C. Dekker, *Room-temperature transistor based on a single carbon nanotubes*, Nature 393 (1998) pp. 49-52.
12. J.M. Planeix, N. Coustel, B. Coq, V. Brotons, P.S. Kumbhar, R. Dutartre, P. Geneste, P. Bernier and P.M Ajayan, *Application of carbon nanotubes as supports in heterogeneous catalysis*, J. Am. Chem. Soc. 116 (1994) pp. 7935-7936.
13. C.N.R. Rao and A.K. Cheetham, *Science and technology of nanomaterials: current status and future prospects*, J. Mater. Chem. 11 (2001) pp. 2887-2894.
14. S. Iijima, *Helical microtubules of graphitic carbon*, Nature 354 (1991) pp. 56-58.
15. H. Zeng, L. Zhu, G. Hao and R. Sheng, *Synthesis of various forms of carbon nanotubes by AC arc discharge*, Carbon 36 (1998) pp. 259-261.
16. Z.J. Shi, Y.F. Lian, X.H. Zhou, Z.N. Gu, Y. Zhang, S. Iijima, L. Zhou and K.T. Yue, *Mass-production of single-wall carbon nanotubes by arc discharge method*, Carbon 37 (1999) pp. 1449-1453.
17. A. Thess, R. Lee, P. Nikolaev, H. Dai, P. Petit, J. Robert, C. Xu, Y.H. Lee, S.G. Kim, A.G. Rinzler, D.T. Colbert, G.E. Scuseria, D. Tombnek and J.E. Fischer, *Crystalline ropes of metallic carbon nanotubes*, Science 273 (1996) pp. 483-487.
18. V.O. Nyamori and N.J. Coville, *Effect of ferrocene/carbon ratio on the size and shape of carbon nanotubes and microspheres*, Organometallics 26 (2007) pp. 4083-4085.
19. D. Jain, A. Winkel and R. Wilhelm, *Solid-state synthesis of well-defined carbon nanocapsules from organometallic precursors*, Small 2 (2006) pp. 752-755.
20. J. Liu, M. Shao, Q. Xie, L. Kong, W. Yu and Y. Qian, *Single-source precursor route to carbon nanotubes at mild temperature*, Carbon 41 (2003) pp. 2101-2104.

21. M. Laskoski, W. Steffen, J.G.M. Morton, M.D. Smith and U.H.F. Bunz, *Synthesis and explosive decomposition of organometallic dehydro[18] annulenes: An access to carbon nanostructures*, J. Am. Chem. Soc. 124 (2002) pp. 13814-13818.
22. B.El Hamaoui, L. Zhi, J. Wu, U. Kolb and K. Müllen, *Uniform carbon and carbon/cobalt nanostructures by solid-state thermolysis of polyphenylene dendrimer/cobalt complexes*, Adv. Mater. 17 (2005) pp. 2957-2960.
23. L. Zhi, T. Gorelik, R. Friedlein, J. Wu, U. Kolb, W.R. Salaneck and K. Müllen, *Solid-state pyrolyses of metal phthalocyanines: A simple approach towards nitrogen-doped CNTs and metal/carbon nanocables*, Small 1 (2005) pp. 798-801.
24. J. Wu, B. El Hamaoui, J. Li, L. Zhi, U. Kolb and K. Müllen, *Solid-state synthesis of "bamboo-like" and straight carbon nanotubes by thermolysis of hexa-peri-hexabenzocoronene-cobalt complexes*, Small 1 (2005) pp. 210-212.
25. S. Liu, X. Tang, Y. Mastai, I. Felner and A. Gedanken, *Preparation and characterization of iron-encapsulating carbon nanotubes and nanoparticles*, J. Mater. Chem.10 (2000) pp. 2502-2506.
26. C. Wu, X. Zhu, L. Ye, C. OuYang, S. Hu, L. Lei, and Y. Xie, *Necklace-like hollow carbon nanospheres from the pentagon-including reactants: Synthesis and electrochemical properties*, Inorg.Chem. 45 (2006) pp. 8543-8550.
27. P.I. Dosa, C. Erben, V.S. Iyer, K.P.C. Vollhardt and I.M. Wasse, *Metal encapsulating carbon nanostructures from oligoalkyne metal complexes*, J. Am. Chem. Soc. 121 (1999) pp. 10430-10431.
28. D.C. Li, L. Dai, S. Huang, A.W.H. Mau and Z.L. Wang, *Structure and growth of aligned carbon nanotube films by pyrolysis*, Chem. Phys. Lett. 316 (2000) pp. 349-355.
29. E. Couteau, K. Hernadi, J.W. Seo, L. Thiên-Nga, Cs. Mikó, R. Gaál and L. Forró, *CVD synthesis of high-purity multiwalled carbon nanotubes using CaCO$_3$ catalyst support for large-scale production*, Chem. Phys. Lett. 378 (2003) pp. 9-17.
30. H. Kathyayini, N. Nagaraju, A. Fonseca and J.B. Nagy, *Catalytic activity of Fe, Co and Fe/Co supported on Ca and Mg oxides, hydroxides and carbonates in the synthesis of carbon nanotubes*, J. Mol. Catal. 223 (2004) pp. 129-136.
31. M.C. Bahome, L.L. Jewell, D. Hildebrandt, D. Glasser and N.J. Coville, *Fischer–Tropsch synthesis over iron catalysts supported on carbon nanotubes*, Appl. Catal. A: General 287 (2005) pp. 60-.
32. J. Cheng, X. Zhang, Z. Luo, F. Liu, Y. Ye, W. Yin, W. Liu and Y. Han, *Carbon nanotube synthesis and parametric study using CaCO$_3$ nanocrystals as catalyst support by CVD*, Mater. Chem. Phys. 95 (2006) pp. 5-11.
33. A. Magrez, J.W. Seo, V.L. Kuznetsov and L. Forró, *Evidence of an equimolar C$_2$H$_2$–CO$_2$ reaction in the synthesis of carbon nanotubes,* Angew. Chem. Int. Ed. 46 (2007) pp. 441-444.
34. S.D. Mhlanga, K.C. Mondal, R. Carter, M.J. Witcomb and N.J. Coville, *The effect of synthesis parameters on the catalytic synthesis of multiwalled carbon nanotubes using Fe-Co/CaCO$_3$ catalysts*, S. Afr. J. Chem. 62 (2009) pp. 67-76.
35. E. Dervishi, Z. Li, A.R. Biris, D. Lupu, S. Trigwell and A.S. Biris, *Morphology of multi-walled carbon nanotubes affected by the thermal stability of the catalyst System,* Chem. Mater. 19 (2007) pp. 179-184.
36. T.C. Schmitt, A.S. Biris, D.W. Miller, A.R. Biris, D. Lupu, S. Trigwell and Z.U. Rahman, *Analysis of effluent gases during the CCVD growth of multi-wall carbon nanotubes from acetylene*, Carbon 44 (2006) pp. 2032-2038.
37. C.H. See and A.T. Harris, *CaCO$_3$ supported Co-Fe catalysts for carbon nanotube synthesis in fluidized bed reactors,* Particle Tech. Fluidization 54 (2008) pp. 657-664.
38. A. Eftekhari, P. Jafarkhani and F. Moztarzadeh, *High-yield synthesis of carbon nanotubes using a water soluble catalyst support in catalytic chemical vapour deposition*, Carbon 44 (2006) pp. 1298-1352.

39. B.H. Liu, J. Ding, Z.Y. Zhong, Z.L. Dong, T. White and J.Y. Lin, *Large-scale preparation of carbon-encapsulated cobalt nanoparticles by the catalytic method*, Chem. Phys. Lett. 358 (2002) pp. 96-102.

40. A. Sazbo, A. Mehn, Z. Konya, A. Fonseca and J.B. Nagy, *"Wash and go": sodium chloride as an easily removable catalyst support for the synthesis of carbon nanotubes*, Phys. Chem. Commun. 6 (2003) pp. 40-41.

41. Z. Li, E. Dervishi, Y. Xu, V. Saini, M. Mahmood, O.D. Oshin, A.R. Biris and A.S. Biris, *Carbon Nanotube Growth on Calcium Carbonate Supported Molybdenum-Transition Bimetal Catalysts*, Catal. Lett. 131 (2009) pp. 356-363.

42. Aldrich Chemical Company, *Handbook of Fine Chemicals*, South Africa, 2007/2008.

43. R.T.K. Baker, G.R. Gadsby, R.B. Thomas and R.J. Waite, *The production and properties of filamentous carbon*, Carbon 13 (1975) pp. 211-214.

44. G.G. Tibbetts, M.L. Lake, K.L. Strong and B.P. Rice, *A review of the fabrication and properties of vapor-grown carbon nanofiber/polymer composites*, Comp. Scie. Technol. 67 (2007) pp. 1709-1718.

45. G.G. Tibbetts, *Carbon fibers produced by pyrolysis of natural gas in stainless steel tubes*, Appl. Phys. Lett. 42 (1983) pp. 666-668.

46. U.M. Graham, A. Dozier, R.A. Khatri, M.C. Bahome, L.L. Jewel, S.D. Mhlanga, N.J. Coville and B.H. Davies, *Carbon nanotubes docking stations: a new concept in catalysis*, Catal. Lett. 129 (2009) pp. 39-45.

47. M. Escobar, M.S. Moreno, R.J. Candal, M.C. Marchi, A. Caso, P.I. Polosecki, G.H. Rubiolo and S. Goyanes, *Synthesis of carbon nanotubes by CVD: Effect of acetylene pressure on nanotubes characteristics*, Appl. Surf. Scie. 254 (2007) pp. 251- 256.

48. T. Luo, L. Chen, K. Bao, W. Yu and Y. Qian, *Solvothermal preparation of amorphous carbon nanotubes and Fe/C coaxial nanocables from sulfur, ferrocene, and benzene*, Carbon 44 (2006) pp. 2844-2848.

49. Y. Xiong, Y. Xie, X. Li and Z. Li, *Production of novel amorphous carbon nanostructures from ferrocene in low-temperature solution*, Carbon 42 (2004) pp. 1447-1453.

50. H. Nishino, R. Nishida, T. Matsui, N. Kawase and I. Mochida, *Growth of amorphous carbon nanotube from poly(tetrafluoroethylene) and ferrous chloride*, Carbon 41 (2003) pp. 2819-2823.

51. T. Zhao, Y. Liu and J. Zhu, *Temperature and catalyst effects on the production of amorphous carbon nanotubes by a modified arc discharge*, Carbon 43 (2005) pp. 2907-2912.

52. Z.C. Kang and Z.L. Wang, *On accretion of nanosize carbon spheres*, J. Phys. Chem. 100 (1996) pp. 5163-5165.

53. K. Hata, D.N. Futaba, K. Mizumo, T. Namai, M. Yumura and S. Iijima, *Water-assisted highly efficient synthesis of impurity-free single-walled carbon nanotubes*, Science 306 (2004) pp. 1362-1364.

54. G. Zhang, D. Mann, L. Zhang, A. Javey, Y. Li, E. Yenilmez, Q.Wang, J.P. McVittie, Y. Nishi, J. Gibbons and H.J. Dai, *Ultra-high-yield growth of vertical single-walled carbon nanotubes: Hidden roles of hydrogen and oxygen*, Proc. Natl. Acad. Sci. 102 (2005) pp. 16141-16145.

55. A.G. Nasibulin, D.P. Brown, P. Queipo, D. Gonzalez, H. Jiang and E.I. Kauppinen, *An essential role of CO$_2$ and H$_2$O during single-walled CNT synthesis from carbon monoxide*, Chem. Phys. Lett. 417 (2006) pp. 179-184.

56. S.D. Mhlanga and N.J. Coville, *A facile procedure to shorten carbon nanotubes*, J. Nanostr. Mater. 10 x (2010) pp. 1-10, doi:10.1166/jnn.2010.2394.

57. U.M. Graham, R.A. Khatri, A. Dozier, N.J. Coville, M.C. Bahome, L.L. Jewell and B.H. Davis, *Renewable FT-Liquids using Fe and Fe-Co catalysts supported on carbon nanotubes with novel catalyst docking stations*, Prepr. Pap. - Am. Chem. Soc., Div. Fuel Chem. 52 (2007) pp. 364-366.
58. Y.Z. Jin, C. Gao, W.K. Hsu, A. Huczko, M. Bustrzejewski, M. Rue, C.Y. Lee, S. Acquah, H. Kroto and D.R.M. Walton, *Large-scale synthesis and characterization of carbon spheres prepared by direct pyrolysis of hydrocarbons*, Carbon 43 (2005) pp. 1944-1953.

NANO-MICROCOMPOSITE AND COMBINED COATINGS ON Ti-Si-N/WC-Co-Cr/STEEL AND Ti-Si-N/$(Cr_3C_2)_{75}$-$(NiCr)_{25}$ BASE: THEIR STRUCTURE AND PROPERTIES

A.D.Pogrebnjak[1,2,*], V.V.Uglov[3], M.V.Il'yashenko[2], V.M.Beresnev[4], A. P. Shpak[8], M.V.Kaverin[1,2], N.K.Erdybaeva[5], Yu A.Kunitskyi[8], Yu.N.Tyurin[6], O.V.Kolisnichenko[6], N.A.Makhmudov[7], A.P.Shypylenko[1,2]

1. Sumy Institute for Surface Modification, P.O.Box 163, 40030 Sumy, Ukraine; e-mail: apogrebnjak@simp.sumy.ua
2. Sumy State University, St.R-Korsakov 2, 40002 Sumy, Ukraine
3. Belarus State University, Minsk, Belarus
4. Science Center for Physics and Technology, Kharkov, Ukraine
5. East-Kazahstan State University of Technology, Ust'-Kamenogorsk, Kazahstan
6. O.E.Paton Welding Institute, NAS of Ukraine, Kiev, Ukraine
7. Samarkand State University, Samarkand, Uzbekistan
8. Institute of Metal Physics G.V.Kurdyumova NAS of Ukraine, Kiev, Ukraine

ABSTRACT

Two types of nano-microcomposite coatings Ti-Si-N/WC-Co-Cr and Ti-Si-N/$(Cr_3C_2)_{75}(NiCr)_{25}$ of 160 to 320μm thickness were manufactured using two deposition technologies: cumulative-detonation and vacuum-arc deposition in HF discharge. The combined coatings restored worn areas of tools and demonstrated high corrosion and wear resistance, increased hardness, elastic modulus, and plasticity index. The composition of top coating changed from Ti = 60at.%, N = 30at.%, and Si ≈ 5at.% to N = 20at.% and Ti – the rest. The first series of coatings indicated the following phases: (Ti, Si)N and TiN for thin coating and WC, W_2C for thick one. The second series indicated (Cr_3Ni_2), pure Cr, and little amount of $Ti_{19}O_{17}$ (in transition region) for thick coating and (Ti, Si)N, TiN for thin one. For the first series, grain sizes reached 25nm, hardness was 38GPa., elastic modulus E=(370 _+32)GPa, and plasticity index H/E=0,11-0,12 For the second series, grain sizes were 15nm, hardness essentially exceeded 42GPa ± 4GPa, elastic modulus E=(425+38)GPa, and plasticity index H/E=0,12-0,13 Corrosion resistance in salt solution and acidic media increased and cylinder-surface friction wear decreased.

INTRODUCTION

Nanocomposite materials as a class of nanomaterials is characterized by a heterogeneous structure, which was formed by practically non-interacting phases with grain dimensions 5 to 35nm [1 – 3]. As a rule, components of such structures are amorphous matrix and inclusions of nanocrystalline phases. These amorphous components agree in the best way with nanocrystalline surfaces providing good adhesion and essentially increasing hardness. Small grain dimensions of the second phase in combination with good strength of intergrain boundaries provide high mechanical properties of such composition materials.

Today, nanomaterials are divided into three classes according to their hardness values: hard nanocomposites of ≥ 20 to 40GPa hardness, superhard of 40 to 80GPa, and ultrahard of ≥ 80GPa [3 – 4]. In addition to protecting functions, chemical and machine building industries need restoration of initial tool dimensions for those tools, which already are functioning in industry. For these purposes, tools are coated with thick coatings, the physical and mechanical properties of which are higher than those of a basic material. Usually, alloys (powders) Ni-Cr-Mo [5], hard alloys WC-Co-Cr [6,8] and Cr_3C_2-Ni, and oxide ceramics Al_2O_3, Al_2O_3-Cr_2O_3 [5,7] are used for such coatings.

In such a way, a combination two layers, for example a thick layer of WC-Co-Cr hard alloy of 100μm, which was formed using cumulative or detonation deposition, and Ti-Si-N thin upper layer (units of a micron) with enhanced physical-mechanical characteristics, which was formed by subsequent condensation, was able to provide higher protecting functions and restore worn surface regions.

The aim of this work was manufacturing of Ti-Si-N- , Ti-Si-N/WC-Co-Cr- , and Ti-Si-N/$(Cr_3C_2)_{75}$ -$(Ni-Cr)_{25}$ – based coatings and investigation of their physical and mechanical properties.

EXPERIMENTAL DETAILS

Polished samples of St.45 (0.45%C, Fe the rest) of 4mm and 20mm diameter were coated using vacuum-arc source with high-frequency discharge. Ti alloyed sintered cathode containing 5 to 10wt.% of Si was deposited using the Bulat 3T-device functioning under 5×10^{-5}Pa vacuum and 100A cathode current.. The sputtering was carried out using two regimes: the standard vacuum-arc method, and HF-regime. A bias potential was applied to the substrate from a HF generator, which produced impulses of convergent oscillations with $\leq$ 1MHz frequency, every impulse duration being 60µs, their repetition frequency – about 10kHz. Due to HF diode effect the value of negative auto bias potential occurring in the substrate amounted to 2 to 3kV at the beginning of impulse (after start of a discharger operation). Coatings of 2 to 3.5µm thickness were deposited to steel substrates of 20 and 30mm diameter, and 3 to 5mm thickness without additional substrate heating. A molecular nitrogen was employed as a reactive gas The first series of rounded steel 3 (0.3wt.% C) samples of 20mm diameter and 4 to 5mm thickness was deposited using cumulative-detonation device CDS-1 of the following parameters: 65mm distance to a nozzle cut, 14mm/s displacement velocity, 5 runs, 12Hz pulse repetition frequency (for WC-Co-Cr). After the deposition, the 160 to 320µm thick coating was melted by a plasma jet (without powder) using eroding W electrode. A melted layer thickness was 45 to 60µm. Then, Ti-Si-N thin coating of about 3µm was deposited over the thick one using the same device Bulat 3T (the method was described above).

For the second series of samples, powder mixture $(Cr_3C_2)_{75}$-$(NiCr)_{25}$ was used, with powder size of 37.8µm.The istance to a nozzle was 70mm, displacement velocity – 4mm/sec, 4 passed, pulse repetition frequency was 12Hz, capacity was C = 200µF, and capacity battery was 3.2kV.

For Bulat 3T, conditions for thin coating deposition remained the same.

For element analysis, we applied the following methods: Rutherford back-scattering of $^4He^+$ of 1.76MeV energy (RBS), scanning electron microscopy (SEM) with EDS (REMMA-103M, Selmi, Ukraine), X-ray diffraction (DRON-3 and Advantage 8, USA).

Hardness and elastic modulus were measured using nanoindentation device Nanoindenter II, MTS System Corporation, Oak Ridge TN (USA) with Berkovich pyramid. Elastic modulus was determined using "load-unload" curves, according to Oliver-Pharr method [14]. Scanning tunneling microscope (STM) of 1nm resolution was used to study the thin layer surface morphology.

RESULTS AND DISCUSSIONS

The cumulative-detonation device functioned under conditions of detonation burning of combustion gaseous mixtures. The device (Figure 1) was constructed of the following sites.

A system of pipelines was used to feed components of combustion gaseous mixtures. Basic difference between cumulative-detonation and detonation devices is that the former realized summary energies of detonation combustion products from several specially designed chambers. Cumulative energy allows high-velocity gas flow with several shock waves providing efficient interaction with powder material. In this way, energy of combustion mixtures is rationally used. The rate and temperature of combustion products depend only on combustion conditions in every chamber. Nozzles function not less than 1000 hours. High burning frequency of 15 to 30Hz is able to provide quasicontinuous coating deposition.

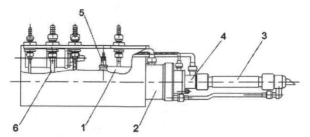

Figure 1 – The cumulative-detonation device.

To increase potential of a technological system, generator allowing one to reach up to 100 kW impulse of high-frequency (HF) discharge was constructed on the basis of pulsed generator with impact contouring. It allowed one to obtain a single pulse of high power under low impedance load, in other words, to operate in a 'shot circuiting' mode.

An advantage of such generators is that their operation weakly depended on changes of loading impedance, which was principally important for functioning under pulsed mode under pulsed loading. To hold up HF discharge, one needed high voltage, which would increase absorption of HF power in discharge. Therefore, in operations with high voltage and high capacity (high current), a special discharge unit would serve as a special commutator.

Discharge started at aluminum coils, then it displaced to copper electrodes having good heat conductivity. The discharge unit had special holes for air cooling of plates. Application of such construction provided high voltages and current and generator operation stability.

Spark resistance of discharge units was calculated using perfect Tamppler formula [13]

$$R = 2.5 \times 10^{-8} \, f/c, \tag{1}$$

where c-is capacity value, which was discharged with f discharge frequency.

Fig.2a shows a scheme for tested generator with f_o=300 kHz, ρ =-10 Ohm, $C= T_{pf}$, L=5 μG, U_o=10 kV, $t_{ipl.}$=30μs, stored capacitor energy was E_o=2,5J, pulsed power $P_{pulse.}$= 8.3 kW. The calculation demonstrated that under periodical discharge $t_{pulse.}$=1.5 μs, P_{pulse} =1.6 MW, repetition frequency f=3.0 kHz , $t_{rep.}$=330 μs, an average power P =7.5 kW, discharge current I=10^3 A.

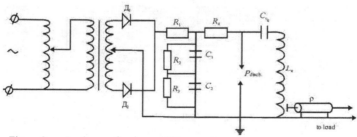

Figure 2a – A scheme of an impact HF generator.

HF generator, which switched plasma-producing antennae, operated in two modes: 1) with an open input (ohmic circuit – zero distance between antenna and ground) and 2) with a closed input –the antenna was separated from the ground by a capacity.

Charged plasma particles in electric HF field of the antennae E reached the following velocity:

$$V = \frac{eE}{mw}, \qquad (2)$$

where e-is a particle charge, m - is particle mass, w -is cyclic generator frequency.
Then, particle range in a field E and under w frequency will be

$$Z_{max} = \frac{eE}{mw^2} \qquad (3)$$

Selecting E and f such that an electron range amounted several centimeter (cm), ion remaining practically non-mobile, the antenna HF voltage started to be detected by a plasma.
Under closed input mode, the antenna was negatively charged, and under open input mode, electrons left the zone of antenna.
Without external magnetic field this zone is:

$$\delta \approx \frac{C}{2\pi \cdot 10^4 \sqrt{n_e}}, \qquad (4)$$

where c- is light velocity; n- is plasma density.

Under action of positive potential ions gained energy for directed motion and bombarded the antenna (under closed input mode) or an internal surface of vacuum chamber and its inside content. This bombardment cleaned the antenna surface, chamber, and tools inside. Ion energy may be controlled by drawing a fraction of charges aside of the antennae by switching higher resistance.
In this case, material sputtering occurred. Maximum voltage amplitude at the beginning of HF impulse was determined by energy value of concrete ions under action of this electric field and by corresponding sputtering efficiency of coated material. The technological device [7, 14] was constructed on the basis of vacuum chamber (7). Grounded metallic walls of vacuum chamber served simultaneously like an anode of vacuum-arc discharge system. Negative potential from an arc-discharge feeding source was applied to a cathode (4), which was fabricated from a material desired for further coating synthesis. A working gas was fed through a gas line (5) using a leak system (1). For additional chemical activation, molecular gases were fed to the vacuum chamber. They passed through a cylindrical quartz discharge chamber (11), in which a generator (12) produced periodically repeated spark discharges. Tools were arranged on a movable table (8). HF voltage was applied to a substrate (8) through the matching device (9) from the HF generator.
In such a way, working with decreasing voltage (Fig.2b) during every impulse, one can join two main technological operations of coating deposition (clearing and deposition), which earlier were performed separately using devices for vacuum-arc deposition. This allowed one to choose better conditions for coating deposition and saved time. Depositing Al_2O_3 and TiN coatings, it was demonstrated that changing HF voltage potential applied to substrate, one could affect coating phase composition [14,15].

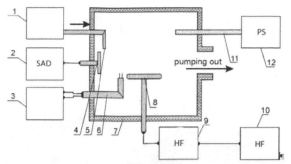

Figure 2a - a scheme of a technological system for coating synthesis operating on the basis of a vacuum-arc discharge: 1 is a device for gas-feeding; 2 – sources for arc-discharge feeding; 3 – measuring probe; 4 – a cathode; 5 – a gas line for gas-feeding; 6 – a double movable probe; 7 – a vacuum chamber; 8 – a substrate; 9 – a device to match a HF-generator; 10 – the HF-generator; 11 – a quartz tube for dissociation of working gas molecules; 12 – power source.

Figure 3 presents an image of nano-microcomposite surface for combined Ti-Si-N/WC-Co-Cr coating.

Figure 3 – Images of surface regions for nano-microcomposite combined coating Ti-Si-N/(Cr$_3$C$_2$)$_{75}$-(NiCr)$_{25}$.

A thin coating was formed using vacuum-arc source and followed the coating surface relief formed by plasma-detonation. Its average roughness varies from 14 to 22μm (after melting and coating deposition using vacuum-arc source). An image of X-ray energy dispersion spectrum is presented below. It indicates the following element concentrations in the thin coating: N ~ 7.0 to 7.52vol.%; Si ~ 0.7vol.%; Ti ~ 76.70 to 81vol.% . For the thick coating we found Fe ~ 0.7vol.%, and traces of Ni and Cr.

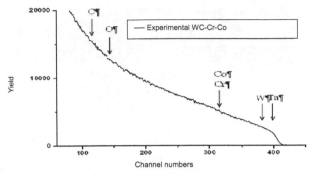

Figure 4a presents RBS data for the thick WC-Co-Cr coating without Ti-Si-N thin one.

Figure 4a – Energy spectra of Rutherford ion backscattering (RBS) for thick coating WC-Co-Cr.

Results for combined coating are presented below, Fig.4b.

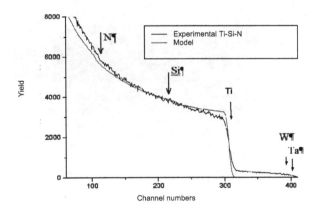

Figure 4b – Energy spectra of Rutherford ion backscattering (RBS) for top thin coating Ti-Si-N/WC-Co-Cr.

Element distribution, which was calculated according to a standard program [5], indicated N = 30at.%; Si ≈ 5 to 6at.%; Ti ≈ 63 to 64at.%. Spectrum of thick coating did not allow us to evaluate element concentration due to high surface roughness of the coating formed by plasma-detonation method.

X-ray analysis of a combined nanocomposite coating is shown in Figure 5.

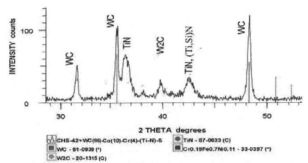

Figure 5. Diffraction patterns fragments obtained for surface region of nano-microcomposite combined coating Ti-Si-N/WC-Co-Cr/ steel substrate.

It indicates the following phases: (Ti, Si)N; TiN – for thin coating, and WC; W_2C – for thick one.

Special samples were prepared for hardness measurements. Their surfaces were grinded and then polished. After grinding, thickness of WC-Co-Cr thick coating decreased to 80 - 90μm. Thin Ti-Si-N film of about 3μm was condensed to the grinded surface. As a result, we found that hardness of different regions essentially varied within 29 ± 4GPa to 32 ± 6GPa. Probably, it is related to non-uniformity of plasma-detonation coating surface, which hardness varied up 11.5 to 17.3GPa . These hardness values remained after condensation of Ti-Si-N thin coating Elastic modulus also features non-ordinary behavior.

Hardness of the thin coating, which was deposited to a polished steel St.45(0.45 % C) surface had maximum value of 48GPa, and its average value H_{av} was 45GPa. Variation of hardness values was lower than that found in a combined coating.

Figure 6 shows dependences of loading-unloading for various indentation depths.

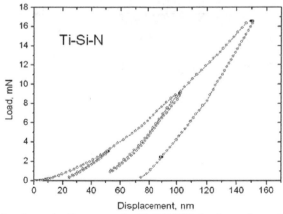

Figure 6 – Loading-unloading curves for Ti-Si-N/WC-Co-Cr coating under various Berkovich indentation depths.

These dependences and calculations, which were performed according to Oliver-Pharr technique [14], indicated that hardness of Ti-Si-N coatings deposited to thick $(Cr_3C_2)_{75}$-$(NiCr)_{25}$ was 37.0 ± 4.0GPa under E = 483GPa.

Figure 7 shows fragments of diffraction patterns for nano-microcomposite combined coating Ti-Si-N/$(Cr_3C_2)_{75}$-$(Ni-Cr)_{25}$.

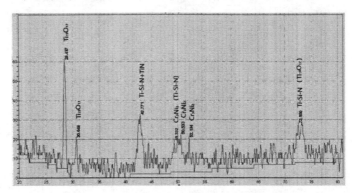

Figure 7 – Diffraction patterns for combined nano-microcomposite coating Ti-Si-N/$(Cr_3C_2)_{75}$-$(NiCr)_{25}$.

These diffraction patterns and calculations of coating structure parameters are presented in Table 1.

Table 1. Calculation results for coating parameters and structures.

N	Angle	Area	Intensity	Half-width	Interplanar	%Max	Phase	hkl
1	28.437	8.511	37	0.4512	3.6416	100.00	Ti_9O_{17}	106 004
2	30.648	3.083	13	0.4518	3.3845	36.84	Ti_9O_{17}	0210 123
3	42.771	10.885	20	1.0490	2.4530	65.79	Ti-Si-N+TiN	111 111
4	49.332	13.862	13	1.9890	2.1433	34.21	Cr_3Ni_2 Ti-Si-N	321 200
5	49.993	17.418	15	2.2322	2.1168	39.47	TiN Ti_9O_{17}	200 1223 110
6	50.533	6.528	12	1.0335	2.0956	44.74	Cr_3Ni_2	330
7	52.134	2.782	15	0.3553	2.0355	39.47	Cr_3Ni_2 Cr Ni	202 110 111
8	72.500	18.056	18	1.8950	1.5127	47.37	Ti_9O_{17} Ti-Si-N	3130 220
9	73.040	11.106	13	1.5950	1.5030	52.63	TiN	220

In the coating, basic phases are Cr_3Ni_2 for the bottom thick coating and (Ti, Si)N and TiN for the thin top coating. Diffraction patterns were taken under cobalt emission. Additionally, we found phases of pure Cr and low concentration of titanium oxide (Ti_9O_{17}) at interphase boundary between thin-thick coatings. Peaks of Ti-Si-N and TiN coincided because of low Si content. (Ti,

Si)N is solid solution based on TiN (Si penetration). The phases are well distinguished at 72 to 73° angles.

Figure 8 shows regions of thick bottom $(Cr_3C)_{75}$-$(NiCr)_{25}$ coating and intensity distribution of X-ray emission for basic elements.

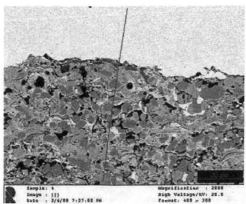

Figure 8a – Regions of transversal cross-section for combined coatings (lines of element analysis are indicated) from SEM and EDS analyses.

In this coating, content of basic elements is the following: nickel and chromium - 36wt.% and 64wt.%, respectively. Also, we found carbon, oxygen, and silicon. Transversal cross-sections did not allow us to distinguish thin upper coating due to its low thickness. We found regions for pure nickel and chromium. Nickel matrix (a white region) indicated high amount of chromium inclusions with various grain dimensions: small grains of < 1μm, average – of 4 to 5μm, and big – of 15 to 20μm.

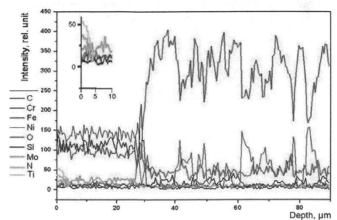

Figure 8b – Element distribution over depth of combined coating Ti-Si-N/$(Cr_3C_2)_{75}$-$(NiCr)_{25}$ for the regions indicated in Fig.8a .

The white region is reach in Ni (to 90at.%). A grey region is reach in Cr (to 92at.%). In these experiments,The inset of Fig.8 shows higher resolution element composition distribution in Ti-Si-N/(Cr$_3$C$_2$)$_{75}$-(NiCr)$_{25}$ nano-microcomposite coating. In a thin nano-composite layer, one can see high Ti concentration, presence of Si, and high enough nitrogen content exceeding 6.7wt.%.

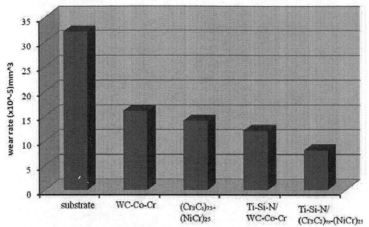

Figure 9 – Histograms of dependences of wear rates for samples, which were fubricated according to scheme cylinder-plane.

Fig. 9 shows results of wear resistance tests, which were performed according to a scheme "cylinder-plane". These results demonstrated the lowest friction wear for Ti-Si-N/(Cr$_3$C$_2$)$_{75}$-(NiCr)$_{25}$ system and the highest friction wear for substrate.

Adhesion between thin Ti-Si-N coating and thick (Cr$_3$C$_2$)$_{75}$-(NiCr)$_{25}$ one was 1.75 times higher than between Ti-Si-N and WC-Co-Cr. In addition, adhesion between thick (Cr$_3$C$_2$)$_{75}$-(NiCr)$_{25}$ coating and steel (substrate) was 7.2 times higher and more than 12.5 times higher than between thick WC-Co-Cr coating and steel (substrate). Maximum adhesion value of about 292N/m was found in the case of Ti-Si-N → (Cr$_3$C$_2$)$_{75}$-(NiCr)$_{25}$.

CONCLUSION

Thick (> 100 μm) nanocomposite coatings of Ti-Si-N/WC-Co-Cr and Ti-Si-N/(Cr$_3$C$_2$)$_{75}$(NiCr)$_{25}$ compositions were formed and investigated.

In the first series of samples, thin coatings contained (Ti, Si)N and TiN phases. Grain dimensions were about 25nm. Hardness reached 38GPa.

In the second series of samples, in thin coating grain dimensions were smaller – about 15nm. Hardness reached 42 to 44GPa. Phase composition was the same – (Ti, Si)N and TiN. Si and N concentrations changed from 10at.% to 5at.% for Si and from 30at.% to 20at.% for N.

Wear resistance increased essentially, which was demonstrated by cylinder-to-sample surface friction. Corrosion resistance and other mechanical characteristics also increased.

ACKNOWLEDGEMENTS

The work was supported by projects ISTCs K-1198 and NAS of Ukraine "Nanosystems, Nanocoatings, Nanomaterils".

The authors acknowledge Drs.S.B.Kislitsyn, Yu.Zh.Tuleushev from Institute for Nuclear Physics, NNC, Almaty, Prof.F.F. Komarov from Belarus State University, Minsk for their help in RBS analysis and corrosion tests.

REFERENCES

1. Pogrebnjak A.D., Shpak A.P., Azarenkov N.A., Beresnev V.M." Structures and Properties of Hard and Superhard Nanocomposite Coatings" // Usp.Phys. **170**, V. 1. 35 – 64. (2009).

2. Musil J.and Zeman P. "Hard a-Si_3N_4/MeNx Nanocomposite Coatings With High Thermal Stability and High Oxidation Resistance" //Solid .State.Phenome, **127**, 31-37 (2007)

3. Zhang R.F.Argon A.S.and Veprek S. "Electronic Structure, Stability, and Mechanism of the Decohesion and Shear of Interfaces in Superhard Nanocomposites and Heterostructures" //Phys.Rev **B79**, 24542.(2009)

4. Veprek S., Veprek-Heijman M.G.J., Karvankova P., Prochazka J.Different approaches to superhard coatings and nanocomposites // Thin Solid Films. V. **476**. P. 1 – 25 (2005)

5. Musil J., Dohnal P., Zeman P. "Physical Properties and High-Temperature Oxidation Resistance of Sputtered Si_3N_4/MoN_x Nanocomposite Coatings" //J. Vac. Sci. Technology. **Vol. 23**, № 4. P. 1568-1574. (2005)

6. Uglov V.V., Anishcik V.M., Zlotski S.V., Abadias G., Dub S.N., "Structural and Mechanical Stability Upon Annealing of Arc-Deposited Ti-Zr-N Coatings" //Surf.and Coat.Tech **202**, 2394-98.(2008)

7. Azarenkov N.A., Beresnev V.M., Pogrebnjak A.D. // Structure and Properties of Coatings and Modified Layers of Materials. Kharkov: (Kharkovskii Nationalnyi Universitet) P. 565. (2007).

8. Nanostructure Coating // Eds. A.Gavaleiro, J.T. De Hosson. (Berlin: Springer-Verlag). P. 340. (2006)

9. Musil J. // Physical and Mechanical Properties of Hard Nanocomposite Films Prepared by Reactive Magnetron Sputtering. Chapter 10 in the Book "Nanostructured hard Coating" A. Cavaleiro, J.Th.M. De Hosson (Eds.). Kluwer Academic/Plenum Publishers. (N.-Y. USA. 2005).

10. Vishniakov Ja.D. // Sovremennye Metody Issledovaniia Struktury Deformirovannykh Kristallov. (Moscow: Metallurgija.) 480.(1975).

11.   Beresnev V.M., Pogrebnjak A.D., Kirik G.V., Edyrbaeva N.K., Ponaryadov V.V. "Structure, and Properties and Fabrication of the Solid Nanocrystalline Coating in Several Ways" // Progress in Physics of Metals. . **Vol. 8., N. 3**. 171 – 246 (2007)

12.   .PogrebnjakA.D, .DanilionokM.M, UglovV.V., ErdybaevaN.K, Kirik,GV.. .DubS.N, .RusakovV.S, .ShypylenkoA.P, ZukovskiP.V., .TuleushevY.Zh "Nanocomposite Protective Coatings Based on Ti-N-Cr/Ni-Cr-B-Si-Fe, Their Structure and properties". //Vacuum. **v.83,** S235-S239 (2009)

13.   Musil.J. Physical and Mechanical Properties of Hard Nanocomposite Films Prepared by Reactive Magnetron Sputtering: invited Chapter 10. In the Book: "Nanostructured Hard Coatings". 2005. Kluwer Academic/Plenum Pullishers. (233 Spring Street, New York, NY 10013, USA. 2007).

14.   KadyrzhanovK. K, KomarovF. F., . PogrebnjakA. D, et al. Ion-Beam Plasma Modification of Materials". (Moscow., Moscow State University(MSU) 2005). 640P.

15. Beresnev V.M.,Tolok V.T.,Shvets O.M..Fursova E.V.,Chernyshev,Malikov L.V. Micro-nanolayers coatings fabricated by vacuum-arc source with discharge//PSE.**v.**4,104-109(2006)

16. New Nanotechniques   eds.A.Malik,R.J.Rawat P.690 Chapter 2.Structure and Properties of Protective Composite Coatings and Modified Surface Prior and After Plasma High Energy Jets Treatment A.D.Pogrebnjak, A.P.Shpak, V.M.Beresnev p.21-115.Nova Science Publ.New-York,2009

PHASE COMPOSITION, THERMAL STABILITY, PHYSICAL AND MECHANICAL PROPERTIES OF SUPERHARD ON BASE Zr-Ti-Si-N NANOCOMPOSITE COATINGS

A.D.Pogrebnjak[1,2,*], O.V.Sobol[3], V.M.Beresnev[4], P.V. Turbin[4], G.V.Kirik[5], N.A.Makhmudov[1,6], M.V Il'yashenko[1],A.P.Shypylenko[1,2] M.V.Kaverin[2], M.Yu.Tashmetov[7], A.V. Pshyk[1,2].

[1.]Sumy Institute for Surface Modification, P.O.Box 163, 40030 Sumy, Ukraine. E-mail: apogrebnjak@simp.sumy.ua
[2.]Sumy State University, 40021 Sumy, Ukraine
[3.]National Technical University, Kharkov, Ukraine
[4.]Science Center for Physics and Technology, Kharkov, Ukraine
[5]. Concern "UKRROSMETAL", Sumy, Ukraine
[6.]Samarkand Branch of Tashkent Information Technology University, Samarkand, Uzbekistan.
[7.]Institute of Nuclear Physics,UAS,Tashkent,Usbekistan

ABSTRACTS
    Zr-Ti -Si-N coating had high thermal stability of phase composition and remained structure state under thermal annealing temperatures reached 1180°C in vacuum and 830°C in air. Effect of isochronous annealing on phase composition, structure, and stress state of Zr-Ti-Si-N- ion-plasma deposited coatings (nanocomposite coatings) was reported. Below 1000°C annealing temperature in vacuum, changing of phase composition is determined by appearing of siliconitride crystallites (B-Si_3N_4) with hexagonal crystalline lattice and by formation of ZrO_2oxide crystallites. Formation of the latter did not result in decay of solid solution (ZrTi)N but increased in it a specific content of Ti-component.
    Vacuum annealing increased sizes of solid solution nanocrystallites from (12 to 15) in as-deposited coatings to 25nm after annealing temperature reached 1180°C. One could also find macro- and microrelaxations, which were accompanied by formation of deformation defects, which values reached 15.5 vol.%.
Under 530°C annealing in vacuum or in air, nanocomposite coating hardness increased. When Ti and Si concentration increased and three phases nc-ZrN, (Zr, Ti)N-nc, and α-Si_3N_4 were formed, average hardness increased to 40,8 ± 4GPa. Annealing to 500°C increased hardness and demonstrated lower spread in values H = 48 ± 6GPa and E = (456 ± 78)GPa.

INTRODUCTION
Recently, nanocomposite coatings of new generation composed of at least two phases with nanocrystalline and/or amorphous structures are of great interest. Due to very small size (10 nm) of their grains and more important role of boundary zones surrounding

single grains, nanocomposite materials behave unlike traditional materials with grain size higher than 100 nm and display quite different properties. Novel unique physical and functional properties of nanocomposites promote rapid development of nanocomposite materials [1-5]. Films with H< 40 GPa and H> 40 GPa hardness are currently described as hard and superhard, correspondingly [5]. As is known, there are two groups of hard and superhard nanocomposites with nc-MeN/hard phase and nc-MeN/soft phase [3-5]. Moreover, bicrystalline phases and/or phases with different crystallographic grain orientations of the same material are distinguished in nanocrystalline and/or amorphous phases. Experimental data of a number of authors demonstrated that Zr-Si-N system was composed of two phases ZrN and (Si, Zr)N [6]. It is possible to assume that Ti addition to this system, would allow one to obtain several phases: nc-ZrN/a-Si_3N_4 and nc-$TiSi_2$ with definite Si and N concentrations. As it is known from works [7,8], properties of solid α – Si_3N_4/McN strongly depend on phase composition and thermal stability of individual phases composing the total coating. It was demonstrated that Zr-Si-N films with ZrN_x (x = 0.8) composition were thermally stable till 1130°C. Those of ZrN_x (x = 1.2) composition, i.e. having higher Zr concentration, like (α – Si_3N_4/MeN) Si_3N_4 + ZrN_x (x = 1.2), were crystallized under higher temperature of 1530°C (x = N/Zr in ZrNx phase).. However, when Si_3N_4 phase was amorphous and took more than 50 vol.% of the coating, hardness ranged from 20 to 40GPa, i.e. did not transit superhardness limit of 40GPa [5,8,9].

We should like to note works [6], in which the authors studied structure stability and mechanical properties of Ti-Zr-N films deposited by vacuum-arc source (Cathodic Arc Vapor Deposition – CAVD) under various plasma densities from metallic cathodes Ti and Zr.

Also we should like to note theoretical works [3,4], which studied electron structure, stability, decohesion mechanism, shear of interfaces in superhard and heterostructures nc-TmN/α-Si_3N_4.

Therefore, the purpose of this work was to study formation of superhard coatings on Zr-Ti- Si-N base and their properties including thermal stability.

EXPERIMENTAL

Coatings were fabricated using vacuum-arc deposition from unit-cast, Zr, Zr-Si, and Zr-Ti-Si targets. Films were deposited in nitrogen atmosphere. Deposition was carried out using standard vacuum-arc and HF discharge methods. Bias potential was applied to substrate from HF generator, which produced impulses of convergent oscillations with $\leq$ 1MHz frequency, every impulse duration being 60μs, their repetition frequency – about 10kHz. Due to HF diode effect, value of negative autobias potential occurring in substrate increased from 2 to 3kV at the beginning of impulse (after start of discharger operation). Coatings of 2 to 3.5μm thickness were deposited to steel substrates (of 20 and 30mm diameter and 3 to 5mm thickness). Deposition was performed without additional substrate heating. Zr-Ti-Si-N coatings were deposited to polycrystalline steel

(St.3 – 0.3wt.%C, Fe the rest). Molecular nitrogen was employed as a reaction gas (Table.1).

Table 1. Physical-technological parameters of deposition

Evaporated materials	Coating	Ia, A	P_N, Pa	URF, V	U, V	Notes
Zr	ZrN	110	0.3	-	200	Standard technology
Zr	ZrN	110	0.3	200	-	HF deposition
Zr-Si	(ZrSi)N	110	0.3	200	-	HF deposition
Ti-Zr-Si	(Ti-Zr-Si)N	110	0.3	200	-	HF deposition

I_a is cathode current in A; P_N is pressure of atomic nitrogen in Pa units; U_{RF} is bias voltage of H_f discharge; U is bias voltage under conditions of vacuum-arc discharge. Annealing was performed in air medium, in a furnace SNOL 8.2/1100 (Kharkov, Ukraine), under temperature T = 300°C, 500°C, and 800°C, and in a vacuum furnace SNVE-1.3, under 5 x 10⁻⁴Pa pressure, and T = 300°C, 500°C, 800°C, and 1180°C.

Studies of phase compositions and structures were performed using X-ray diffraction devices DRON-3M, under filtered emission Cu-K_α, using secondary beam of a graphite monochromator. Diffraction spectra were taken point-by-point, with a scanning step 2Θ = 0.05 to 0.1°.

To study stressed states of the coatings, we applied X-ray strain and stresses measurements ("$a - \sin^2 \psi$" method) and its modifications, which were used to films with a strong texture of axial type. Element compositions were studied using X-ray fluorescent spectrometer SPRUT (AO UkrRoentgen, Ukraine) with a shoot-through tube employing a silver anode, and under exciting voltage 40kV. Surface morphology, structure, and element compositions were analyzed using a scanning electron microscope (REMMA-103M, Quanta-1000) with microanalysis (EDS- energy disperse X-ray spectroscopy). Additionally, to study element composition and stoichiometry, we used RBS under 1.35MeV $^4He^+$ ion energy, 170° scattering angle, and 16keV detector resolution.

Studies of mechanical characteristics were realized with the help of nanoindentation under 10nN load of NANOINDENTOR II (MTS System Inc., USA) indentation device with diamond Berkovich pyramid [9].

EXPERIMENTAL RESULTS AND DISCUSSION
THE COATINGS PREPARED BY SPUTTERING FROM THE Zr-Si-Ti TARGET

Figure 1 shows energy spectra of ion backscattering measured for steel samples with deposited Zr-Ti-Si-N coatings. Since Zr and Ti concentration was high, these spectra could hardly help to determine Si and N background concentration. Measurements of Si and N concentration using eating away of the RBS spectra gave higher error than for Zr and Ti. But still, Si concentration was not less than 7at.%, while that of N might reach more than 15at.%.

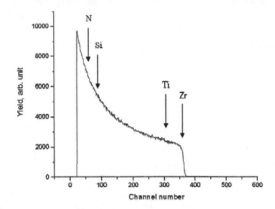

Figure. 1. Energy spectrum of 1.35MeV He$^+$ ion Rutherford back-scattering measured for Zr-Ti-Si-N nanocomposite coating; arrows indicate kinematical boundaries of elements.

Analyzing phase composition of Zr-Ti-Si-N films, we found that a basic crystalline component of as-deposition on state was solid solution (Zr, Ti)N based on cubic lattice of structured NaCl. In Table 2, we presents x-ray diffraction curves: a lattice period in non-stressed cross-section (a_0), value of macrodeformation ε, microdeformation $<\varepsilon>$, and concentration of packing defects $\alpha_{def.pack}$. The data were obtained both for samples after coating deposition and for those annealed in vacuum and air under various temperatures.

Table 2. Changes of structure and substructure parameters occurring in ion-plasma deposited films of Zr-Ti-Si-N system in the course of high-temperature annealing in vacuum and in air.

Parameters of structure	After deposi-tion	T_{an}=300 °C vacuum	T_{an}=500 °C vacuum	T_{an}=800 °C vacuum	T_{an}=1100 °C vacuum	T_a=300 °C air	T_{an}=500 °C air
a_0, nm	0,45520	0,45226	0,45149	0,45120	0,45064	0,45315	0,45195
ε, %	-2,93	-2,40	-1,82	-1,01	-1,09	-2,15	-1,55
$<\varepsilon>$, %	1,4	1,0	0,85	0,5	0,8	0,95	0,88
$\alpha_{def. pack.}$	0,057	0,085	0,107	0,155	0,150	0,090	0,128

Crystallites of solid (Zr, Ti)N solution underwent compressing elastic macrostresses occurring in a "film-substrate" system. Compressing stresses, which were present in a plane of growing film, indicated development of compressing deformation in a crystal lattice, which was identified by a shift of diffraction lines in the process of angular surveys ("$\sin^2\psi$ – method") and reached – 2.93% value (Table 2). With $E \approx 400$GPa characteristic elastic modulus and 0.28 Poisson coefficient, deformation value corresponded to that occurring under action of compressing stresses $\sigma_c \approx -8.5$GPa. We should also note that such high stresses characterize nitride films, which were formed under deposition with high radiation factor, which provided high adhesion to base material and development of compression stresses in the film, which was stiffly bound to the base material due to "atomic peening"- effect.

At substructure level, microdeformation was still high, and amounted 1.4%. With a relatively small average crystallite size ($L \approx 15$nm), development of such high microdeformation indicated significant contribution of crystallite deformed boundaries.

Phase composition of ion-plasma films under temperature of vacuum annealing lower than 1000°C remained practically unchanged, corresponding to post as-deposition state. An average crystallite size of solid solution (Zr, Ti)N also remained practically unchanged. Under this temperature range (300-1000°), microdeformation at substructure level typically decreased from 1,4 to 0.8% (Table 2), which indicated decreasing amount of lattice defects.

Compressing macrodeformation partially relaxed when annealing temperature increased within 25 to 1000°C range. Practically, it decreased by a factor of three, reaching a value $\varepsilon \approx -1.1\%$ under $T_{an} = 1000$°C. We should note that $\varepsilon \approx -1\%$, which was close to that obtained under annealing, was reached in the case of pure, ordered ZrN ion-plasma deposited coatings. A lattice period a_0 defined for non-stressed cross-section (under $\sin^2\psi_0 = 0.43$) decreased with decreasing annealing temperature (Table 2). If one would relate such decreased period to ordering of titanium atoms with lower atomic radius, which were built-in into metallic sublattice instead of Zr atoms, then using Vegard's rule, the decrease from 0.4552nm to 0.4512nm corresponded to 8.5at.% to 19.5at.%.increase of titanium atom content.

Shift of diffraction lines to various directions corresponding to planes taken at θ- 2θ (according to Bregg-Brentano scheme) seems to be explained by packing defects, which are present in metallic fcc-sublattice. Concentration of packing defects may be evaluated by comparison of shifting (222) and non-shifting (333) peak positions[10]. After condensation, average packing defect concentration in a lattice of (Zr, Ti)N solid solution was 5.7%. As a result of annealing, packing defect concentration increased and reached 15.5% under $T_{an} = 800$°C.

Qualitative changing of phase composition was observed in films under vacuum annealing at $T_{an} > 1000$°C. Figure 2a shows characteristic diffraction curve, which was taken under 30min annealing at $T_{an} = 1100$°C. Under high-temperature annealing, in addition to (Zr, Ti)N nitrides (which period was close to ZrN lattice) and (Ti, Z)N (which period was close to TiN lattice), we observed diffraction peaks from zirconium oxide crystallites (ZrO_2, according to JCPDS Powder Diffraction Cards, international Center for Diffraction Data 42-1164, hexagonal lattice) and titanium oxide (TiO, JCPDS 43-1296, cubic lattice), and, probably, initial amorphous β-Si_3N_4 phase crystallites (JCPDS 33-1160, hexagonal lattice). Appearance of zirconium and titanium oxides was related to oxidation relaxation under coating surface interaction with oxygen atoms coming from residual vacuum atmosphere under annealing. Under annealing temperatures below 1000°C, coatings phase composition remained practically unchanged (Fig.2 a,b,). One could note only changed width of diffraction lines and their shift to higher diffraction angles. The latter characterizes relaxation of compressing stresses in coatings. Changed diffraction lines were related to increased crystalline sizes (in general) and decreased micro-deformation

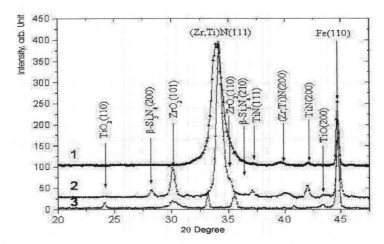

Figure 2a. Region of X-ray diffraction spectra taken for the coatings of Zr-Ti-Si-N system after deposition (1); after 30 min annealing in vacuum, under T_{an} = 500°C (2), and under T_{an} = 800°C in air (3). Three peaks, which are not designated in the curve, are for an oxide of Fe_2O_3 substrate (JCPDS 33-0664).

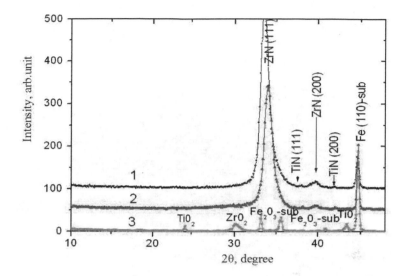

Figure 2b. XRD diffraction patterns for Zr-Ti-Si target in 0.3Pa nitrogen atmosphere (vacuum-arc source with HF discharge): 1 - for initial (as-received) samples; 2 - for annealed at 500°C (30min in vacuum); 3 - for annealed at 800°C (30min in vacuum).

Figure 3a shows the film cross-section, which demonstrates that in the course of deposition, no cracks were found, that indicated good quality of the coating. Figure 3b shows chemical composition over coating cross-section. Spectra indicate that N concentration (for second series) changed from 3.16 to 4.22wt.%, Si concentration was about 0.98 to 1.03wt.%, Ti was 11.78 to 13.52 wt%. and that Zr = 73.90 to 77.91wt.%. These results indicated that amount of N is essentially high, and this allowed it to participate in formation of nitrides with Zr , Ti, or (Zr, Ti)N solid solution. Si concentration was low, however, results reported by Veprek et al. [3, 4] indicated Si concentration as high as 6 to 7at.%, which was enough to form siliconitride phases.

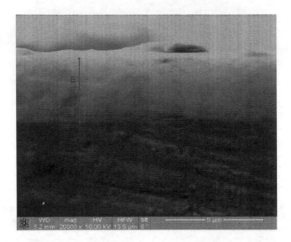

Figure 3a. Cross-section of hard coating Zr-Ti-Si-N (high concentration of Si and N)

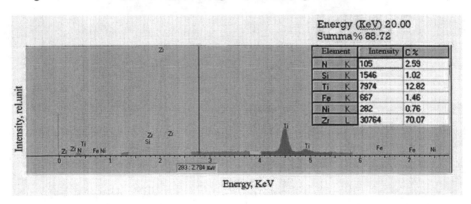

Figure 3b. Data of microanalysis for some point of Zr-Ti-Si-N (Ti≈12%) nanocomposite coating surface.

Changes occurred under macrodeformation of crystallites of basic film phase – (Zr, Ti)N solid solution. Compressing deformation of crystallite lattices increased, which seemed to be related to additional new crystalline components, which appeared in film material: oxides and siliconitrides. In the lattice itself, a period decreased corresponding to increased Ti concentration. Ordered atoms in metallic (Zr/Ti) sublattice of solid solution increased from 8.5 to 21at.%.

In this temperature range, crystallite size increased from 15 to 25nm, crystallite lattice microdeformation increasing non-essentially up 0.5 to 0.8%. Table 2 summarizes substructure characteristics of $(Zr, Ti)N$ solid solution crystallites.

Figure 4a, shows XRD-diffraction patterns, and lower (b), a histogram of volume phases for nano-structured Zr-Si-N coating with 10 to 12nm grain sizes for *nc-ZxH* phase (where *nc* is a nano-structured phase). These data demonstrate 17% volume fraction of quasi-amorphous α-Si_3N_4 phase, 54% of nano-composite nano-structured phases, and the rest was α-Fe from samples substrates.

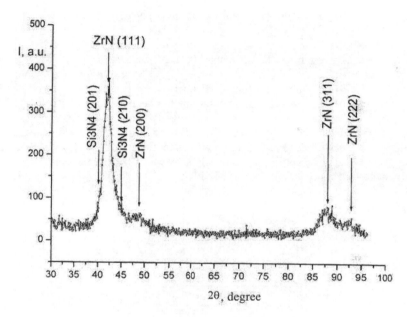

Figure 4a. A fragment of diffraction patterns for Zr-Si-N coating deposited by vacuum-arc method with HF stimulation (Fe-K_α radiation).

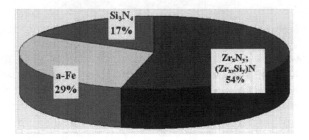

Figure 4b. Histogram of phase ratio for nanocomposite coatings Zr-Si-N

Figure 5 shows nano-hardness vs annealing temperatures for coatings fabricated using vacuum-arc source with HF discharge (hardness values for Zr-Ti-N and Zr-Si-N systems are presented for comparison with Zr-Ti-Si-N coating). We should like to note that "self-hardening effect" was observed for those nano-structured coatings, which were deposited to substrates under low temperatures (not exceeding 120°C to 150°C). Therefore, the process of spinodal segregation along nano-grain boundaries was not terminated because of energy deficit for diffusion rate. When annealing temperature came close to 550°C to 600°C range, the process of spinodal segregation was over, i.e. all nano-grains were totally surrounded by an interlayer of several α-Si_3N_4 nano-layers (quasi-amorphous phase).

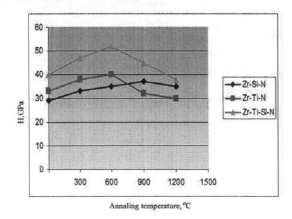

Figure 5.Nano-hardness vs annealing temperatures for coatings fabricated using vacuum-arc source with HF discharge (hardness values for Zr-Ti-N and Zr-Si-N systems are presented for comparison with Zr-Ti-Si-N coating

In initial state, after deposition, those samples (second series), which phase composition included three phases (Zr,Ti)N-nc, ZrN-nc, and α-Si$_3$N$_4$), hardness was H = 40,6 ± 4GPa; E = 392 ± 26GPa (Fig. 5). 500°C annealing increased H and E and decreased spread in hardness values, for example, H = 48 ± 6GPa and E = (456 ± 78GPa).

In such a way, hardness, which was increased in the process of annealing, seems to be related to incomplete spinodal phase segregation at grain boundaries resulting from deposition of Zr-Ti-Si-N- (nanocomposite).

Annealing stimulated spinodal phase segregation [3,4], forming more stable modulated film structures with alternating in volume concentration of phase components (ZrN; (Zr,Ti)N; Si$_3$N$_4$).

CONCLUSION

In such a way, decreased concentration of active oxygen atoms coming from annealing atmosphere increased stability of film phase composition from 500 to 1000°C. Changing crystalline phase composition was determined by crystallization of siliconitrides and formation of β − Si$_3$N$_4$ crystallites with hexagonal lattice, as well as low ZrO$_2$ concentration formed in the film surface.

High macro- and microdeformation occurring in the coating seems to be related to an "atomic peening" effect resulting to non-ordered distribution of titanium atoms implanted to the film during its growth. In the course of annealing, the macro- and micro-deformation relaxed. The relaxation was accompanied by formation of deformation packing defects in a metallic sublattice of (Zr, Ti)N solid solution. This can be revealed by X-ray scanning, which demonstrated shift and broadening of diffraction peaks. Highest content of packing defects indicated shift of most closely packed planes in a fcc-sublattice (111) with respect to each other [1,12] and became pronounced under vacuum annealing at T$_{an}$ = 800 to 1100°C reaching 15.5vol.%.

When Ti and Si concentration increased (second series) and three phases nc-Zr-N, (Zr, Ti)N-nc, and α-Si$_3$N$_4$ were formed, average hardness increased to 40,8 ± 4GPa. Figure 6 shows that in initial as-deposition state, Zr-Ti-Si-N film (as received) had 40.8 GPa hnanohardness. After annealing (a dark dotted curve) at 500°C in vacuum, coating nanohardness reached H = 55.3GPa. (see, fig.5)

In Zr-Ti-Si-N coatings, increased Ti concentration, formation of three phases- (Zr, Ti)N-nc-57vol.%, TiN-nc-35vol.%, and α − Si$_3$N$_4$ ≥ 7.5 vol.%, as well as changes of grain size, which decreased to (6 to 8)nm in (Zr, Ti)N and (10 to 12) in TiN in comparison with first series resulted in increased nanohardness and decreased difference in hardness values. Annealing in vacuum below 500°C finished the process of spinodal segregation at grain boundaries and interfaces. Annealing stimulated segregation processes and formed stable modulated coating structure [1,4,8].

The work was funded by the program "Nanosystems, Nanomaterials and Nanocoatings. New Principles in Nanomaterial Manufacturing by Ion, Plasma and Electron Beams" NAS of Ukraine.

REFERENCES

1. Pogrebnjak A.D., Shpak A.P., Azarenkov N.A., Beresnev V.M." Structures and Properties of Hard and Superhard Nanocomposite Coatings" // Usp.Phys. **170**, V. 1. 35 – 64. (2009).
2. Musil J.and Zeman P. "Hard a-Si3N4/MeNx Nanocomposite Coatings With High Thermal Stability and High Oxidation Resistance" //Solid .State.Phenome, **127**, 31-37 (2007)
3. Zhang R.F.Argon A.S.and Veprek S. "Electronic Structure, Stability, and Mechanism of the Decohesion and Shear of Interfaces in Superhard Nanocomposites and Heterostructures" //Phys.Rev **B79**, 24542.(2009)
4. Veprek S., Veprek-Heijman M.G.J., Karvankova P., Prochazka J. Different approaches to superhard coatings and nanocomposites // Thin Solid Films. V. **476**. P. 1 – 25 (2005)
5. Musil J., Dohnal P., Zeman P. "Physical Properties and High-Temperature Oxidation Resistance of Sputtered Si_3N_4/MoN_x Nanocomposite Coatings" //J. Vac. Sci. Technology. **Vol. 23**, № 4. P. 1568-1574. (2005)
6. Uglov V.V., Anishchik V.M., Zlotski S.V., Abadias G., Dub S.N., "Structural and Mechanical Stability Upon Annealing of Arc-Deposited Ti-Zr-N Coatings" //Surf.and Coat.Tech **202**, 2394-98.(2008)
7. Azarenkov N.A., Beresnev V.M., Pogrebnjak A.D. // Structure and Properties of Coatings and Modified Layers of Materials. Kharkov: (Kharkovskii Nationalnyi Universitet) P. 565.(2007).
8. Nanostructure Coating // Eds. A.Gavaleiro, J.T. De Hosson. (Berlin: Springer-Verlag). P. 340. (2006)
9. Musil J. // Physical and Mechanical Properties of Hard Nanocomposite Films Prepared by Reactive Magnetron Sputtering. Chapter 10 in the Book "Nanostructured hard Coating" A. Cavaleiro, J.Th.M. De Hosson (Eds.). Kluiver Academic/Plenum Publishers. (N.-Y. USA. 2005).
10. Vishniakov Ja.D. // Sovremennye Metody Issledovaniia Struktury Deformirovannykh Kristallov. (Moscow: Metallurgija.) 480.(1975).
11. Beresnev V.M., Pogrebnjak A.D., Kirik G.V., Edyrbaeva N.K., Ponaryadov V.V. "Structure, and Properties and Fabrication of the Solid Nanocrystalline Coating in Several Ways" // Progress in Physics of Metals. . **Vol. 8., N. 3**. 171 – 246 (2007)
12. A.D.Pogrebnjak, M.M.Danilionok, V.V.Uglov, N.K.Erdybaeva, G.V.Kirik, S.N.Dub, V.S.Rusakov, A.P.Shypylenko, P.V.Zukovski, Y.Zh.Tuleushev "Nanocomposite Protective Coatings Based on Ti-N-Cr/Ni-Cr-B-Si-Fe, Their Structure and properties". //Vacuum. **v.83,** S235-S239 (2009).

CHARACTERIZATION OF NANOCRYSTALLINE SURFACE LAYER IN LOW CARBON STEEL INDUCED BY SURFACE RAPID MULTI-ROLLING TREATMENT

Chang Sun[a,c,d], Kangning Sun[a,b,*] Jingde Zhang[a,b] ,Pengfei Chuan[a,b], Xiaoxin Wang[a,b] , Nan Wang[a,b]

[a] Key Laboratory of Liquid Structure and Heredity of Materials, Ministry of Education, Shandong University(south part), Jingshi Road 17923, Jinan 250061, PR China
[b] Engineering Ceramics Key Laboratory of Shandong Province, Shandong University(south part), Jingshi Road 17923, Jinan 250061, PR China
[c] School of Mechanical Engineering, Shandong University(south part), Jingshi Road 17923, Jinan 250061, PR China
[d] Shandong Supervision and Inspection Institute for Product Quality, Shandabeilu Road 81, Jinan 250100, PR China

ABSTRACT

A nanostructured surface layer was formed on low carbon steel by means of surface rapid multi-rolling treatment (SRMT). The microstructure of the surface produced by SRMT was systematically characterized by optical microscopy (OM), X-ray diffraction (XRD) and Field emission scanning electron microscopy (FE-SEM). Microhardness tester was used in order to examine the hardness variation by different treatment conditions. The results showed that the microstructure of the surface layer of the low carbon steel is refined to nanoscale by means of SRMT. Deformed layer of about 20μm in thickness is formed under the conditions of 1800 rps, 0.4 MPa and 45 minutes. The mean grain size in the top surface layer is approximately 30 nm. The hardness of nanostructured surface layer was enhanced significantly after SRMT compared with that of the initial sample.

INTRODUCTION

Nanocrystalline materials, which are characterized by a microstructural length or grain size of up to about 100 nm, posses unique physical, chemical and mechanical properties[1]. In the past decades, many researchers have paid more attention to nanocrystalline materials, and developed a series of methods to synthesize them, such as inter gas condensation[2-3], crystallization of amorphous solid[4], ball-milling[5], electrodeposition[6], and severe plastic deformation[7], etc. In many cases, materials failure originate from surface, for example high surface stress, stress concentration, fatigue, external corrosion, temperature change, and wear, etc. Therefore, surface nanocrystallization can greatly enhance the mechanical properties and corrosion resistance to aggressive environment without changing the chemical constituents[8-9]. Recently, surface severe plastic deformation has attracted great scientific interests. By means of it, the surface nanocrystallization can be attained in most metallic materials. In recent years there are several methods to produce surface nanocrystallization by means of surface severe plastic deformation, such as surface mechanical attrition treatment (SMAT)[10-13], ultrasonic shot peening (USSP)[14-15], high energy shot peening (HESP)[16-18], circulation rolling deformation (CRPD)[19-20], wire-brushing[21], etc. On the base of the principle of surface severe plastic deformation, we present a new method for surface nanocrystallization, and we

*Corresponding author.Tel.:+86-531-88392439;fax: +86-531-88392439
E-mail: sunkangning@sdu.edu.cn

named it surface rapid multi-rolling treatment (SRMT).

In this work, a low carbon steel was selected to be treated by surface rapid multi-rolling treatment (SRMT), and A nanostructured surface layer was formed on low carbon steel. The microstructure of the surface produced by SRMT was systematically characterized by optical microscopy (OM), X-ray diffraction (XRD) and Field emission scanning electron microscopy (FE-SEM). Microhardness tester was used in order to examine the hardness variation by different treatment conditions.

EXPERIMENT

Fig. 1 shows structure principle of SRMT device. A number of alloy steel balls inlaying on rotating disk roll rapid on sample. The process of the multi-rolling of sample is as follows: static pressure 0.4 MPa, rotating disk rate 1800r/min feeding rate 1m/min. Alloy steel balls distribute in according to continuous helix arrangement as shown in fig. 2. The photo of SRMT device is also given in fig.3.

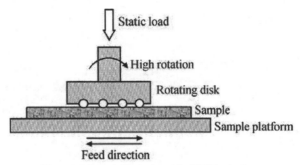

Figure 1. Schematic illustration of the SRMT technology

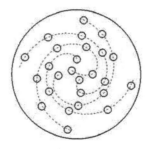

Figure 2. Orientation arrangement of the rolling balls on the rotating disk

Figure 3. The photo of SRMT device

The materials used in this study were a commercial Q235 low carbon steel. The samples were treated by SRMT for 30, 45, 60 min respectively. The microstructure evolution of samples was characterized by optical microscopy and Hitachi SU-70 field-emission SEM. X-ray diffraction (XRD) analysis of the surface layer in the as-treated sample was carried out on a Rigaku D/MAX-rA X-ray diffractometer with CuK_α. The average grain size was obtained from XRD line broadening in means of the Scherrer equation. The microhardness on the top surface was measured by a HX-1000T microhardness meter.

RESULTS AND DISCUSSION
1. XRD analysis

Fig. 4 shows the XRD patterns of the samples after SRMT for different time. From the XRD spectra we can see that there is not second-phase, and diffraction peaks are considerably broadened and the broadening becomes wider as the processing duration turns much longer. The average grain size was calculated in the terms of the diffraction line broadening by the Scherrer formula. The results listed in Table 1. From table 1 we can see that the grain size reach nanoscale about 27 nm at 30 min treatment. With the increase of duration, the grain size decreases. But when the process duration reach up to 60 min, the grain size does not decrease.

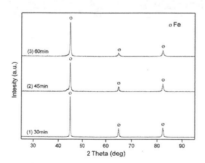

Figure 4. XRD pattern of surface nanocrystalline sample treated for different time

Table I. Average grain size in the surface layer of the treated samples calculated from XRD data

Processing duration (min)	Average grain size (nm)
30	27.8
45	23.8
60	24.4

2. Optical microscope observation

Fig. 5 shows the optical micrographs of the samples observed in the cross section. As shown in fig.5-d, it was found that basic structure of original samples was pearlitic structure and ferrite matrix. and the average size is 40-100µm. Fig. 3-a, b and c show the optical micrograph within the surface nanocrystallization after SRMT at different processing duration. They showed the geomorphology of the deformed layers. The plastic deformation is in homogeneous along the depth. The depth of the deformation is about 100µm . The thickness the severe plastic deformation layer is about 40µm During the SRMT, surface of samples was impacted at high strain rate and rolling, and severe plastic deformation was induced into materials.

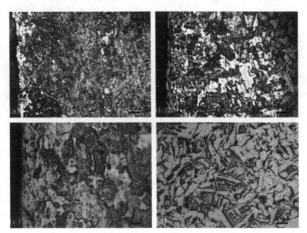

Figure 5. Optical morphologies of the cross-section (a) SRMT for 30min, (a) SRMT for 45min, (a) SRMT for 60min and (d) original sample

3. FE-SEM observation

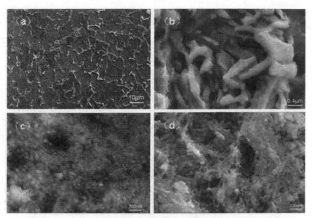

Figure 6. FE-SEM image of the top surface layer of the SRMT sample (a)untreated, (b) treated for
30min, (c) treated for 45min and (d) treated for 60min

The nanocrystalline surface layer of the sample was observed by Field emission scanning electron
microscopy (FE-SEM). The results are shown in fig. 6. In panel b, we are looking at the grain size,
which were refined after SRMT for 30 min compared with non-treated samples (panel b), but the
surface is not smooth. In panel c and d, the grain size is less than that of the sample treated for 30 min.
From the image, the mean grain size could be measured. The average size on the top surface layer of
the sample treated for 45 min is about 50 nm. The average grain size determined by FE-SEM
observation is evidently bigger than that obtained from the XRD calculation. This is perhaps as a result
that some smaller grains on the top surface were etching out before the FE-SEM sample was observed.
Fig. 7 shows the cross-section of the sample at about 10μm deep from the top surface treated for 45
min. It is clear that the grains are flake-shaped and appear as a regular stacked structure which is
possibly caused by rolling continuously at same direction.

Figure 7. FE-SEM image of the cross-section of the SRMT sample treated for 45 min

4. Measurement of microhardness

Fig. 8 shows that the microhardness variation with the processing duration. It can be seen that, with the increase of treatment time, the microhardness of the low carbon steel increases significantly in the top surface layer, which can be obviously attributed to grain refinement and work-hardening. The grain boundary in the surface nanocrystallization layer can block the dislocation movement rendering the material to have higher microhardness. This result is consistent with the results obtained by FE-SEM observation.

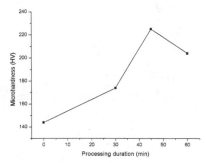

Figure 8. Microhardness of nanocrystalline surface of low carbon steel after SRMT for different time

CONCLUSION

In this paper it is shown that a nanocrystalline surface layer on low carbon steel can be prepared by means of the surface rapid multi-rolling treatment (SRMT). The grain size is about 30 nm in the top surface layer and increases with depth. The increase of SRMT duration does not change singificently the grain size of top layer. The microhardness of the nanocrystalline surface layers is improved significantly after SRMT compared with that of the untreated sample. The microhardness is about HV 225 at the top surface, while the value of the unprocessed sample is about HV 144. We can see that the surface hardening effect is obtained obviously. In the sub surface layer of 8-12μm, the grain size is about 100-500nm. The grain refinement mechanism might be the one most relevant to the stack of deformation and the staggered shear.

ACKNOWLEDGEMENTS

This work was supported by the Natural Science Foundation of the People's Republic of China (Grant No. 30540061), China Postdoctoral Science Foundation (No. 200804401138) and Shandong Province Postdoctoral Innovation Foundation (No. 200902030).

FOOTNOTES
*Corresponding author. Tel.:+86-531-88392439; fax: +86-531-88392439; E-mail: sunkangning@sdu.edu.cn

REFERENCES
[1]   Gleiter H. Nanocrystalline materials [J]. Progress in Materials Science, 1989,33(4): 223-315.

[2] Hai NH, Lemoine R, Remboldt S, et al. Iron and Cobalt-based magnetic fluids produced by inert gas condensation[J]. Journal of Magnetism and Magnetic Materials, 2005,293(1): 75-79.

[3] Suryanarayana C, Prabhu B. Synthesis of Nanostructured Materials by Inert-Gas Condensation Methods[M]//CARL C K. Nanostructured Materials (Second Edition). Norwich, NY; William Andrew Publishing. 2007: 47-90.

[4] Becker C, Conrad E, Dogan P, et al. Solid-phase crystallization of amorphous silicon on ZnO:Al for thin-film solar cells[J]. Solar Energy Materials and Solar Cells, 2009,93(6-7): 855-858.

[5] Koch CC. Synthesis of nanostructured materials by mechanical milling: problems and opportunities[J]. Nanostructured Materials, 1997,9(1-8): 13-22.

[6] Sealy C. Nanowire fabrication redefined from top to bottom: Fabrication and processing[J]. Materials Today, 2006,9(12): 13-13.

[7] Villegas JC, Shaw LL. Nanocrystallization process and mechanism in a nickel alloy subjected to surface severe plastic deformation[J]. Acta Materialia, 2009,57(19): 5782-5795

[8] Ya M, Xing Y, Dai F, et al. Study of residual stress in surface nanostructured AISI 316L stainless steel using two mechanical methods[J]. Surface and Coatings Technology, 2003,168(2-3): 148-155.

[9] Wang T, Yu J, Dong B. surface nanocrystallization induced by shot peening and its effect on corrosion resistance of 1Cr18NiTi stainless steel[J]. Surface and Coatings Technology, 2006,200: 4777-4781.

[10] Wang ZB, Lu J, Lu K. Wear and corrosion properties of a low carbon steel processed by means of SMAT followed by lower temperature chromizing treatment[J]. Surface and Coatings Technology, 2006,201(6): 2796-2801.

[11] Guo FA, Ji YL, Zhang YN, et al. Local thermal property analysis by scanning thermal microscope of ultrafine-grained surface layer in copper and in titanium produced by surface mechanical attrition treatment[J]. Materials Characterization, 2007,58(7): 658-665.

[12] Hu T, Wen CS, Lu J, et al. Surface mechanical attrition treatment induced phase transformation behavior in NiTi shape memory alloy[J]. Journal of Alloys and Compounds, 2009,482(1-2): 298-301.

[13] Wen M, Gu J-F, Liu G, et al. Formation of nanoporous titania on bulk titanium by hybrid surface mechanical attrition treatment[J]. Surface and Coatings Technology, 2007,201(14): 6285-6289.

[14] Tao NR, Sui ML, Lu J, et al. Surface nanocrystallization of iron induced by ultrasonic shot peening[J]. Nanostructured Materials, 1999,11(4): 433-440.

[15] Wu X, Tao N, Hong Y, et al. Microstructure and evolution of mechanically-induced ultrafine grain in surface layer of AL-alloy subjected to USSP[J]. Acta Materialia, 2002,50(8): 2075-2084.

[16] Ni Z, Wang X, Wang J, et al. Characterization of the phase transformation in a nanostructured surface layer of 304 stainless steel induced by high-energy shot peening[J]. Physica B: Condensed Matter, 2003,334(1-2): 221-228.

[17] Hou L-f, Wei Y-h, Liu B-s, et al. Microstructure evolution of AZ91D induced by high energy shot peening[J]. Transactions of Nonferrous Metals Society of China, 2008,18(5): 1053-1057.

[18] G L, C WS, F LX, et al. Low carbon steel with nanostructured surface layer induced by high-energy shot peening[J]. Scripta Materials, 2001,44: 1791-1795.

[19] Hui-qiong Y, Xin-min F. Surface nanocrystalline of 7A04 aluminium alloy induced by circulation rolling plastic deformation[J]. Transactions of Nonferrous Metals Society of China, 2006,16: s656-s660.

[20] Xin-min F, Bo-sen Z, Lin Z, et al. Surface nanocrystalline of low carbon steel induced by circulation rolling plastic deformation[J]. material science forum, 2005,(475-479): 133-136.

[21] Sato M, Tsuji N, Minamino Y, et al. Formation of nanocrystalline surface layers in various metallic materials by near surface severe plastic deformation[J]. Science and Technology of Advanced Materials, 5(1-2): 145-152.

PROPERTIES OF NANO-METAL CARBIDE CONTAINED Mg-TiC (SiC) COMPOSITES

Mustafa Aydin[a]* and Rasit Koc[a]
[a]Southern Illinois University at Carbondale, Department of Mechanical Engineering and Energy
Processes Carbondale, IL 62901, USA

ABSTRACT
This paper deals with the processing of nano-size metal carbide containing Mg/nano-TiC (SiC) composites and their properties. In this study, nano-size composites were produced using high energy ball milling. The particle size of the TiC and SiC nano particles used as reinforcement particles were 50 nm and 600 nm, respectively. The composites obtained on the Mg and Mg/n-xwt%TiC (SiC) (x: 1, 5, 10) compositions were investigated by X-ray diffraction (XRD), optic, scanning electron microscopy (SEM), Archimedes principle, and microhardness tests. The results of microstructural characterization obtained from Mg and Mg/n-wt%TiC (SiC) composites also revealed a near-defect free interface formed between matrix and reinforcement particles. An increase in the weight percentage of nano-SiC and nano-TiC reinforcement lead to an increase of hardness in the both SiC and TiC reinforced composites. It was shown that the nano-size TiC(SiC) particles, at the fracture surface of all of samples, were homogeneously distributed in the matrix. The nano-sized particles were located in the grain boundaries of magnesium matrix. Compared to Mg samples, ductility of composites increased with increasing amount of TiC and SiC nano particles.

INTRODUCTION
In the present, Magnesium (Mg) is one of the lightest industrials and structural metals available (1-4). Mg and its alloys are the most effective materials for reducing the weight of components. Mg and its alloys are gaining considerable attention in recent years due to the increasing demand of light weight structural materials in automotive and aerospace industries. The density of magnesium is less than 35.6% and 61.3% that of aluminum and titanium, respectively (5-7). However, industrials application of Mg and its alloys are often restricted due to their inherent deficiencies such as low stiffness, high wear rate, high chemical reactivity and low mechanical strength at high temperature. But these drawbacks can be overcome by adding ceramic reinforcing particles (1, 8). Further enhancement in mechanical properties can be achieved by adding or forming nanoparticles in magnesium matrix (8). The mechanical properties of pure Mg composites can be further enhanced by decreasing the grain sizes of ceramic particulates and/or matrix grain sizes from micrometer to nanometer (2). Mg based nano composites can even exhibit higher tensile strength and ductility than their microcomposite counter parts. This is especially important for structural applications when high strength and ductility are essential. So the improvement of the mechanical and microstructural properties is very important for the increase of application areas.

*Corresponding author, [a]Mustafa AYDIN, (Present address) Southern Illinois University at Carbondale, Department of Mechanical Engineering and Energy Process, Carbondale, IL 62901 USA
Tel: +1 618 453 3680, Fax: +1 618 453 7658, E-mail: maydin@engr.siu.edu
* (Permanent address) Dumlupinar University, Department of Mechanical Engineering, Main campus, 43100, Kutahya, TURKEY, Tel: +902742652031 E-mail: m_aydin@dumlupinar.edu.tr

In recent years, producing of discontinuously micro size powders with reinforced magnesium composites has been applied by different conventional fabrication methods, such as mechanical stir casting, squeeze casting and powder metallurgy (9). Although the stir casting and squeeze casting methods are easy and cheap, they have some problems such as how to distribute and disperse of nano-scale particle uniformly into the matrix materials. One of the major problems in fabrication of nanosize ceramic powders such as TiC, SiC, TiN, AlN is the existence of agglomerates in fine particles mainly due to high specific surface area (10). However in the high energy ball milling, which is one of the powder metallurgy techniques, the reinforcement particle can be successfully distributed in the matrix. So in this study, the powder metallurgy technique is selected to distribute and disperse nanoparticles into magnesium matrix.

Accordingly, the primary aim of the present study was to fabrication and investigation of the properties of Mg/n-SiC and Mg/n-TiC composites by using powder metallurgy method. Particular emphasis was placed to study the effect of the presence of nano particles such as SiC and TiC particulates and increasing the content at the final properties of pure magnesium.

EXPERIMENTAL PROCEDURES

In this study, pure Mg powders (>99.8%, Alfa Aesar, -325 micron, USA), SiC and TiC powders were used as the starting materials. The SiC powders were supplied by Merck Company, German and were 600 nm sizes. As the TiC nano particles, the powders patented by US005417952A were used (11). These powders have 50 nm sizes and produced by precursor and coating methods (12, 13).

At the production steps of Mg/n-SiC and Mg/n-TiC composites, they were used by powder metallurgy route. And the composites samples were fabricated at Mg/n-xwt%SiC and Mg/n-xwt%TiC (x; 1, 5, 10) compositions. During the productions of Mg composites, the same steps were used as in previous study of Aydin et al. (8). Pure Mg, nanosized silicon carbide and titanium carbide powders were carefully weighted, loaded into a vial cup and mixed under the argon atmosphere. All these processes of powders were done in a glove-box to minimize the exposure of moisture and contamination, which would be detrimental to the mechanical properties of the composite parts (14). The weighted powders were then mixed in a Turbola Spex 8000 high planetary ball mill at 400 rpm for 30 min in argon with 1:5 ball ratio. After the mixing, the powders was compacted under a pressure of 550 MPa to fabricate billets of 31 mm diameter and 5 mm height (for XRD sample), 13 mm diameter and 4 mm height (optic and SEM samples). After compaction, the green compacted samples were sintered at a temperature of 580°C, at duration of 1 h with 4°C/min heating ratio into the tube furnace under the argon atmosphere. After milling, a small quantity of powder was collected for examination of structure changes by means of X-ray diffraction (XRD). In addition, sintered samples were examined to check for structure changes. The phase analysis was studied on the polished samples of Mg, Mg/n-SiC and Mg/n-TiC using an automated Rigaku mini flex II diffractometer. Thin samples (13 mm diameter and 4mm thickness) were exposed to Cu Kα radiation (λ=1.5418A°) with a scan speed of 2 deg min^{-1}. The diffraction range was 20°-80° with a step size of 0.05°. Experimental density measurements were performed in accordance with Archimedes' principle (15) on polished samples of Mg and Mg/n-SiC and Mg/n-TiC. Distilled water was used as the immersion liquid. All the samples were weighed using a Mettler Toledo, Newclassic ML 104 electronic balance, with an accuracy of 0.0001g. Theoretical densities of materials were calculated assuming they were full-dense and there was no reinforcement/matrix interfacial reaction to measure the volume percentage of porosity in the end materials (15) and for composites samples, the rule-of-mixture was used. Microhardness measurements were made on the polished Mg and Mg/n-SiC, Mg/n-TiC composites samples. Vickers microhardness was measured by Schimadzu 2000 MxT 50 automatic digital microhardness tester using 200 gf-intenting loads for 10 sec. The fracture surface characterization studies were carried out on the fractured Mg and Mg/SiC-Mg/TiC nano composite samples in order to provide of fracture mechanism. Fracture surface characterization studies were primarily accomplished by using Hitachi S-570 scanning electron microscopy (SEM).

RESULTS AND DISCUSSIONS
X-ray diffraction analysis
    Fig.1a and 1b show the XRD spectra of milled powders and sintered composite samples. All Mg/n-SiC and Mg/n-TiC samples show that only Mg, SiC and TiC peaks were detected. After one hour sintering, the samples still contain only the peaks of elemental powders such as Mg, Si, Ti and C. As it can be seen, all of the production steps such as weighting, mixing, pressing and sintering were selected under proper circumstances. These composite samples were not contaminated with oxygen during the balancing, mixing, pressing and sintering. As expected, when the ratio of the reinforcement particle was increased, the intensity of the peaks increased.

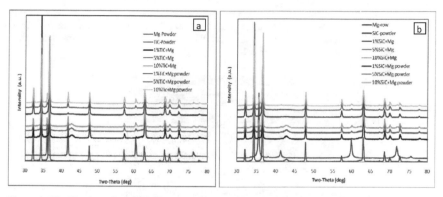

Figure 1. The X-ray analysis of before and after sintering of a) pure Mg, SiC and milled–sintered Mg/n-SiC composites, and b) pure Mg, TiC and milled–sintered Mg/n-TiC composites.

Density
    The porosity results of Mg and composite materials were given in Table 1. According to Table 1, the measured porosity values for all the samples without 10% SiC and 10%TiC were found to be very low. These porosity results correspond with the results of other authors (14, 15, 17). The results of density measurement show that 1-5%SiC and 1-5%TiC composite samples have higher density than 10%SiC and 10%TiC composites (see Table 1). After the 5% reinforcement ratio, the porosity increased. As Ugandher and Gupta (18) stated that the porosity of Mg-4.8%SiC and Mg-15.4%SiC composites was measured as 0.53 and 1.98, respectively. In this study, the highest porosity was exhibited by Mg/n-10wt%TiC samples. Comparatively, the 10wt%TiC sample displayed highest porosity level than Mg/n-10wt%SiC samples. Because the particle size of the TiC powders was smaller than the SiC nano particles sizes, TiC powders were agglomerated grain boundary of Mg matrix. Hence, the porosity of TiC reinforced composite samples was increased. As a reason of this, it was figured out that the samples have closed porous like cell occurred between the grain boundaries of Mg composites.

Table 1. The changes of density and porosity of Mg, Mg/n-SiC and Mg/n-TiC composites

Material	Reinforcement ratio, (wt%)	Theoretical density, (g/cm³)	Experimental density, (g/cm³)	Porosity, (%)
Pure Mg	0	1.7400	1.7396 ± 0.0002	0.023 ± 0.01
Mg/n-%SiC	1	1.7547	1.7539 ± 0.0003	0.046 ± 0.02
	5	1.8135	1.8007 ± 0.0002	0.710 ± 0.01
	10	1.8870	1.8232 ± 0.0002	3.382 ± 0.01
Mg/n-%TiC	1	1.7718	1.7685 ± 0.0002	0.187 ± 0.01
	5	1.8990	1.8762 ± 0.0002	1.201 ± 0.02
	10	2.0580	1.8794 ± 0.0002	8.680 ± 0.01

Microstructures

Fig. 2 and Fig. 3 show optical microstructure of the transversal section of the pure Mg, and Mg/n-SiC and Mg/n-TiC composites for the different reinforcement ratio. As it can be seen, the pure Mg (unmilled) particles showed approximate grain diameters at about 30-40 μm (Fig. 2). The Mg/n-SiC and Mg/n-TiC composites showed median matrix grain diameters at about 10-30 μm according to scale (Fig. 3a, b, c, d, e, f). The size of pure magnesium particles decreased according to starting particle size, because during the milling process, brittle Mg powder crushed with a hard steel ball, a vial cup surface and a reinforcement particle such as TiC or SiC.

Figure 2. The optical microstructures of pure Mg sample after pressing and sintering.

In these microstructures, the gray areas show Mg matrix and dark areas show reinforcement particles such as SiC (Fig. 3a,b,c) and TiC (Fig. 3d,e,f). The dark areas are microcluster of SiC and TiC nanoparticles. As it can be seen, these dark areas have more 10%TiC than the 10%SiC, because the size of TiC nano particles are smaller than the SiC nano particles. The clusters were distributed uniformly into the matrix. When the reinforcement particle ratio was increased, the dark area ratio of composites increased from 1% to 10%. In these microstructures, the reinforcement particles were distributed as homogeneous and in general, the particles were located in the grain boundaries of matrix Mg particles. However, few ceramic nanoparticles also disperse into the grains of magnesium matrix. Hence, it is expected that the hardness of the composite parts will be increased. Optical analysis revealed that the composites materials were not contaminated by impurities such as iron from vial cup, from the milling ball or from the air (19).

Fig.4 shows that the SEM images the morphology and distribution of SiC and TiC nanocomposites in the Mg matrix together with higher magnification. In the SEM images, it can be seen that the nanoparticles are well distributed and dispersed into the matrix (9). It should be noted that these microstructures were focused on the SiC and TiC microclusters, so that the grain boundaries of the Mg matrix could be seen clearly. In these Figures, the assembled white areas are SiC and TiC nanoparticles and these clusters were distributed uniformly (4). Fortunately SiC and TiC nanoparticles were well dispersed into the Mg matrix, the significant improvement of mechanical properties would mostly be attributed. The grain size views revealed that all the composite samples exhibited considerably lower grain size when compared to original pure magnesium (15). The grain boundaries of Mg matrix were surrounded by SiC and TiC particles as networks and these networks were distributed as homogenous in the matrix. In a study by Eugene and Gupta (20), it was reported that the nano-sized SiC particles were decorated as a continuous network at the Mg particles boundaries, and was similar to that observed by Ferkel and Modrike (19).

Figure 3. The optical microstructures of a)Mg/n-1%SiC, b) Mg/n-5%SiC, c) Mg/n-10%SiC, d) Mg/n-1%TiC, e) Mg/n-5%TiC, and f) Mg/n-10%TiC composite samples.

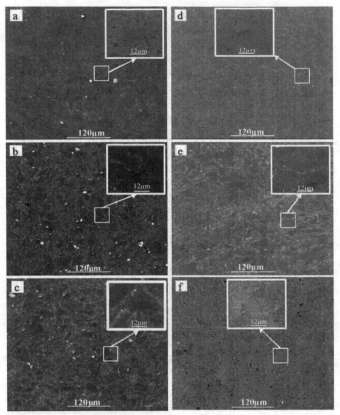

Figure 4. The SEM images of a)Mg/n-1%SiC, b)Mg/n-1%TiC, c)Mg/n-5%SiC, d)Mg/n-5%TiC, e)Mg/n-10%SiC, f)Mg/n-10%TiC composite parts with higher magnification.

As it can be seen in Fig.3 and 4, the microstructures of Mg/n-SiC, Mg/n-TiC composites and their high magnification were found to be uniformly distributed within matrix for all the composites. This indicates that the processing parameters which are weighting, mixing, pressing and sintering are capable of producing composites of homogeneous distribution of reinforcements in magnesium matrix. The uniform distribution of reinforcements is critical for obtaining good mechanical properties of composites. The quality of interfacial integrity could not be determined due to limitations of the SEM machine (14). In general, the observation also unveiled minimal presence of porosity in the composite material (17). When the Mg matrix was reinforced with nano sized TiC powders at 10% ratio, as seen in the Table 1 and Fig.4, the porosity of Mg/n-10%TiC composite increased when compared to 10%SiC. The porous of 10%TiC composite samples were indicated on the SEM image in Fig.4e. Figure 4 indicates that the duration of 30 min. mixing time for this study is not enough for homogenous mixing of nano-sized particle reinforced composite materials.

The SiC and TiC microclusters were confirmed by energy dispersive spectroscopy (EDS) in Figure 5. In Fig. 5a shows the EDS analysis of Mg/n-SiC composite parts and SiC clusters and the Si peak can be seen clearly. Fig. 5b shows the EDS analysis of Mg/n-TiC composite parts and TiC clusters and the Ti peak can be seen clearly. But in the EDS of the pure Mg matrix, no Si, Ti or other impurity elements peak can be seen. As a result, during the fabrication steps of composite materials, the samples were not contaminated by the vial cup, and other contaminants.

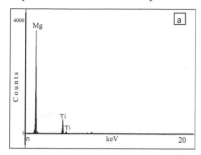 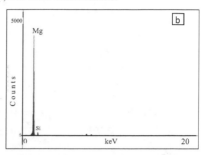

Figure 5. The EDS analysis of a)Mg/n-5wt%TiC, and b)Mg/n-5wt%SiC composite parts.

Microhardness

The microhardness of the pure Mg and Mg/n-SiC and Mg/n-TiC composite samples are given in Figure 6. The results of microhardness measurement explained that the addition of nano-size reinforcement leads to a significant increase in both Mg/n-SiC and Mg/n-TiC nanocomposite samples. An increase in the weight percentage of nano-SiC and nano-TiC reinforcement lead to an increase of hardness in the both SiC and TiC reinforced composite samples. In Figure 6, pure Mg samples display the lowest microhardness value when compared with all composite samples. There was a tendency of decreasing microhardness with a decrease in amount of SiC-TiC nano particles. Because TiC powders have a smaller size than that the SiC powders, Mg/n-10%TiC composite sample showed the highest microhardness value. The increase of hardness can be attributed to presence of harder n-SiC, n-TiC particles and surface bonding strength of particles and reduced grain size (see Fig.2). In addition, dispersion hardening will probably play a role in hardening of the composite parts. Because some of the n-SiC and n-TiC particles were found within the Mg grains even when they were still located close to the Mg matrix grain boundaries. In the literature, the hardness of the Mg composites reinforced with different reinforcement particles such as 1.10%Al$_2$O$_3$ (21), 5%AlN (1), 4.8-15.3SiC (18), 7.3-14%Ni (22) were measured as 51-70, 60-83, 53-56, 69-82 HV, respectively. In this study, the microhardness of the Mg/n-SiC and Mg/n-TiC composites were corresponded with above studies and the hardness of pure Mg was increased from 46 to 78 HV.

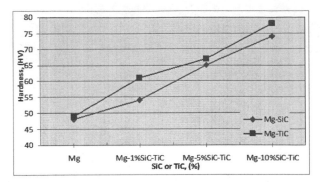

Figure 6. The microhardness of the pure Mg and Mg/n-SiC and Mg/n-TiC composite samples

Fracture surfaces

Generally, the samples fractured at room temperature exhibited a mixed mode of brittle cleavage fracture with dimple-like ductile fracture. As it can be seen in Figure 7 and 8a, the results of fracture analysis indicated typical brittle fracture in the case of pure Mg samples. This can be attributed to the HCP crystal structure of Mg that restricts the slip to the basal plane (22, 1). In Figure 8b and 8c, the fracture behavior of matrix was changed from brittle to ductile in the case of Mg/n-1%TiC(SiC) and Mg/n-5%TiC(SiC) composites for pure Mg due to the incorporation of nano particles such as TiC and SiC. Compared to Mg samples, ductility of composites increased with increasing amount of TiC and SiC nano particles. In Fig.7b and 7c, the dimples occurred on the fracture surface which can be seen easily and these dimples of 10%SiC composite are smaller than the 5%SiC composite sample. Compared to 1%TiC(SiC) and 5%TiC(SiC) samples, ductility of composites reduced with increasing amount of 10%TiC(SiC). When different metal or ceramic particles such as Ti, Mo, CNT and nano-Al_2O_3 were added to pure Mg, the ductility of Mg composites were increased and these results have also been observed by this author (20), too. As a result, the tests also showed the capability of nanosized carbide particulates reinforcement to increase the ductility of pure Mg. The increase in ductility of Mg matrix due to presence of nano particles can primarily be attributed to the presence of uniformly distributed reinforcement particulates. The hardest particles dispersed in brittle matrix act as a ductility enhancer which dislocation movement is restricted (15). In the presence of porosity in 10%TiC composite material, also supported by the experimental density value and porosity ratio (see Table 1).

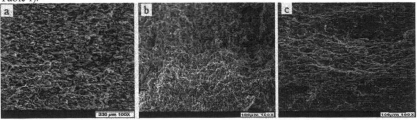

Figure 7. Fracture surfaces of the samples, a) pure Mg, b) Mg/n-5wt%SiC, c) Mg/n-10wt%TiC

In Fig.8, the fracture surfaces of the pure Mg and Mg composites reinforced with n-SiC and n-TiC were shown in higher magnifications than the magnification shown in Fig.7. In Fig.8, the highest magnification of SEM images for fracture surfaces given into higher magnification. Fig.8a shows that the fracture surface of pure Mg and occurred dimples are lower than the shown in Fig.8b and 8c. At the highest magnification of grain boundary of Mg/n-10%TiC composites are more nano-sized TiC particles than n-SiC particles. Additionally to in higher magnification of Mg/n-10%TiC, the porous are clearly shown in grain boundaries. In the composite materials, the fractures occurred in these areas. It may be noted that the presence of a higher ratio in the grain boundaries lead to the plastic incapability, thus it serves as a crack center leading to the reduction in ductility.

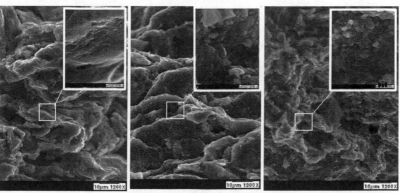

Figure 8. Fracture surfaces of the samples for higher magnification, a) pure Mg, b) Mg/n-10%SiC, c) Mg/n-10%TiC

CONCLUSION

The production of Mg and Mg/n-SiC, Mg/n-TiC composite materials were successfully accomplished using the high energy ball milling method. Important features such as minimal oxidation and contamination of magnesium using glove-box and argon, absence of macropores and porosity for Mg/1-5%n-TiC(SiC) composites revealed in this study. During the production steps, the any reaction was not detected between Mg-SiC and Mg-TiC powders. The results of microstructural characterization, obtained from Mg and composites materials, also revealed a near–defect free interface which is formed between reinforcement and Mg matrix. The hardness of Mg considerably increased when it was reinforced with hard nano-size particles such as SiC and TiC. In this study, the hardness of pure Mg was increased from 46 to 78 HV. The observation of optic and SEM results showed that approximately all of nano particles were homogeneously distributed and particles were located in the grain boundaries of Mg matrix. It may be noted that the presence of a higher ratio (10%SiC, 10%TiC) in the grain boundaries of nano particles lead to the plastic incapability, thus it serves as a crack center leading to the reduction in ductility.

ACKNOWLEDGMENTS

The authors would like to thank TUBITAK (The Science and Technological Research Council of Turkey) for financial support during the course of this investigation.

REFERENCES

[1]M. A. Thei, L. Lu, M. O. Lai, Mechanical properties of nanostructural Mg-5wt%Al-xwt%AlN composite synthesized from Mg chips, *Composite Structures*, **75**, 206-212 (2006).

[2]C. S. Tjong, Novel nanoparticle-reinforced metal matrix composites with enhanced mechanical properties, *Advanced Engineering Materials*, DOI:10.1002/adem.200700106.

[3]M. Gupta, M.O. Lai, Sarvanaranganathan D., Synthesis, microstructure and properties characterization of disintegrated melt deposited Mg/SiC composites, *Journal of Materials Science* **35**, 2155-2165 (2000).

[4]X. Li, H. Konishi, G. Cao, Mechanical properties and microstructure of Mg/SiC nanocomposites fabricated by ultrasonic cavitation based nanomanufacturing, *Journal of Manufacturing Science and Engineering*, **130**, 31105-1-6 (2008).

[5]L. Lu, K.K. Thang, M. Gupta, M-based composite reinforced by Mg₂Si, *Composites Science and Technology*, **63**, 627-632 (2003).

[6]M. O. Lai, L. Lu, W. Laing, Formation of magnesium nanocomposite via mechanical milling, *Composite Structure*, **66**, 301-304 (2004).

[7]C.S. Goh, M. Gupta, J. Wei, L.C. Lee, Characterization of high performance Mg/MgO nanocomposites, *Journal of Composite Materials*, **41**, 2325 (2007).

[8]M. Aydin, C. Ozgur, O. San, Microstructure and hardness of Mg-based composites reinforced with Mg2Si particles, *Rare Metals*, **28**, 4, 396-400 (2009).

[9]J. Lan, Y. Yang, X. Li, Microstructure and microhardness of SiC nanoparticles reinforced magnesium composites fabricated by ultrasonic method, *Materials Science and Engineering A* **386**, 284-290 (2004).

[10]R.G. Reddy, Processing of nanoscale materials, *Reviews on Advanced Materials Science*, **5**, 121-133 (2003).

[11]United state patent, US005417952A, Koc R., Glatzmaier G.C., Patent number; 5,417,952, Date of patent; May 23 1995.

[12]R. Koc, C. Meng, G.A. Swift, Sintering properties of submicron TiC powders from carbon coated titania precursor, *Journal of Materials Science*, **35**, 3131-41 (2000).

[13]R. Koc, J.S. Folmer, Synthesis of submicrometer titanium carbide powders, *Journal of American Ceramic Society*, **80** , 4, 953-956 (1997).

[14]S.K. Thakur, G.T. Kwee, M. Gupta, Development and characterization of magnesium composites containing nano-sized silicon carbide and carbon nanotubes as hybrid reinforcement, *Journal of materials science*, **42**, 10040-46 (2007).

[15]S.F. Hassan, M. Gupta, Effect of different types of nano-sized oxide particulates on microstructural and mechanical properties of elemental Mg, *Journal of Materials Science*, **41**, 2229-36 (2006).

[16]H.Y. Wang, Q.C. Jiang, X.L. Li, J.G. Wang, Q.F. Guan, H.Q. Liang, In situ synthesis of TiC from nanopowders in a molten magnesium alloy, *Materials Research Bulletin*, **38**, 1387-92 (2003).

[17]S.F. Hassan, M. Gupta, Enhancing physical and mechanical properties of Mg using nanosized Al2O3 particulates as reinforcement, *Metallurgical and Materials Transactions A*, **36**A, 2253-58 (2005).

[18]S. Ugandhar, M. Gupta, S.K. Sinha, Enhancing strength and ductility of Mg/SiC composites using recrystallization heat treatment, *Composite Structures*, **72**, 266-272 (206).

[19]M. Ferkel, B.L. Mordike, Magnesium strengthened by SiC nanoparticles, *Materials Science and Engineering A*, **298**, 193-199 (2001).

[20]W.W.L. Eugene, M. Gupta, Simultaneously improving strength and ductility of magnesium using nano-size SiC particulates and microwaves, *Advanced Engineering Materials*, **8**, 735-39 (2006).

[21]M. Gupta, S.F. Hassan, Development of high performance magnesium nano-composites using nano-Al₂O₃ as reinforcement, *Materials Science and Engineering A*, **392**, 163-168 (2005).

[22]S.F. Hassan, M. Gupta, Development of high strength magnesium based composites using elemental nickel particulates as reinforcement, *Journal of Materials Science*, **37**, 2467-74 (2002).

Author Index